Introduction to Global Plate Tectonics V:
Australia & Antarctica

By
William A. Szary

Library of Congress Cataloging In Publication Data

Szary, William A.
Introduction to Global Plate Tectonics V. Australia & Antarctica.
William A. Szary.

Includes references.

ISBN-13: 978-1505678192
ISBN-10: 1505678196

1.Plate tectonics; 2. Paleogeography; 3. Historical Geology; 4.
General Geology; 5. Earth Sciences.

Printed by CreateSpace, an Amazon Company

Earth2Energy Educational Publishing
Tampa FL 33618
Earth2Energy is a Registered Trademark

Table of Contents

Chapter I: Australia

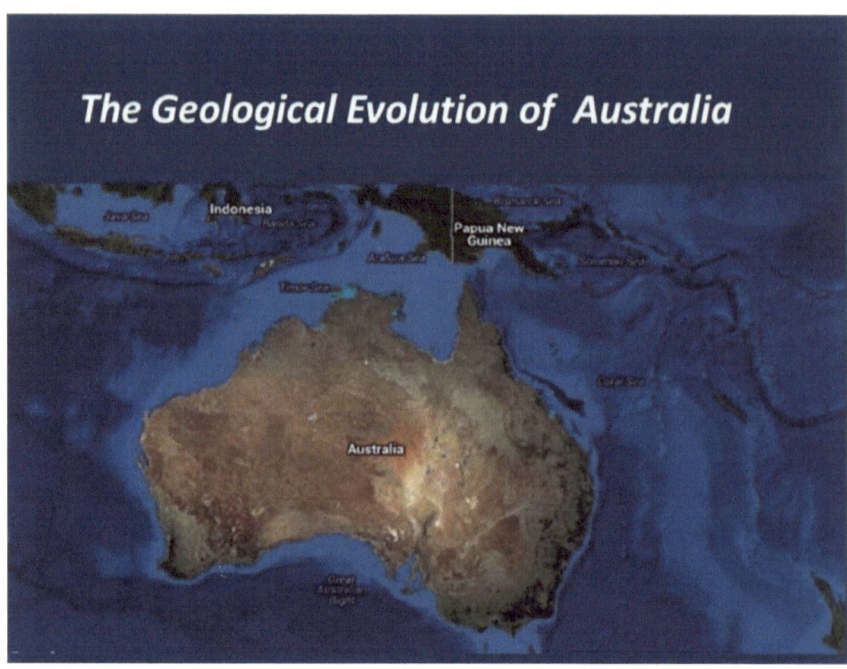

The Australian continent grew in size from west to east. Early Precambrian rocks formed the basement in the west. Middle Precambrian rocks formed the central region, and Phanerozoic aged rocks formed the eastern part of the continent.

Three Precambrian Era cratons make up the continent: West Australia, North Australia, and South Australia. Precambrian continental growth was tied to the geologic histories of Kenorland, Nuna (Columbia), Rodinia, and Pangea-Gondwana.

Four time periods transpired during the tectonic evolution of Australia: 3.8 to 2.2 billion years ago; 2.2 to 1.3 billion years ago; 1.3 billion to 700 million years ago; and, 700 million years ago to the present.

Cratonic nuclei grew during the first time period. The three later periods included the breakup and scattering of land masses when Nuna, Rodinia, and Pangea-Gondwana fragmented.

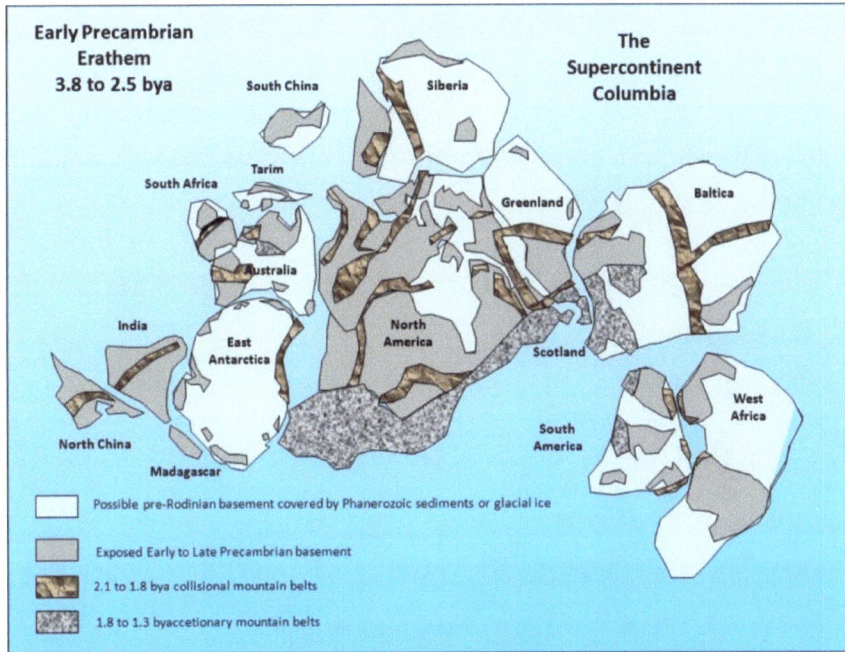

Figure 1. Australia-Antarctica and India were accreted together along the western Laurentian continent during the existence of the supercontinent called Nuna (Columbia).

When the supercontinent Nuna (also called Columbia) existed, Australia was positioned in the center west part of the supercontinent, attached to the northern margin of Antarctica (**Figure 1**). South Africa was attached to northwest Australia.

The two major Australian cratons (gray shading) were called the Yilgarn and Pilbara terranes. These two terranes made up Western Australia. Rocks consisted of anorthosite and orthogneiss.

Anorthosites are igneous rocks that are injected into continental crust, forming massifs and batholiths when exposed at the surface by erosion of the overlying crust. Orthogneiss is a metamorphic rock derived from other igneous rocks.

The mineral zircon was found in sedimentary rocks in the Jack Hills area, age dated at 4.4 billion years ago, suggesting the Narryer Terrane in the Yilgarn craton were much older than 2.8 billion years.

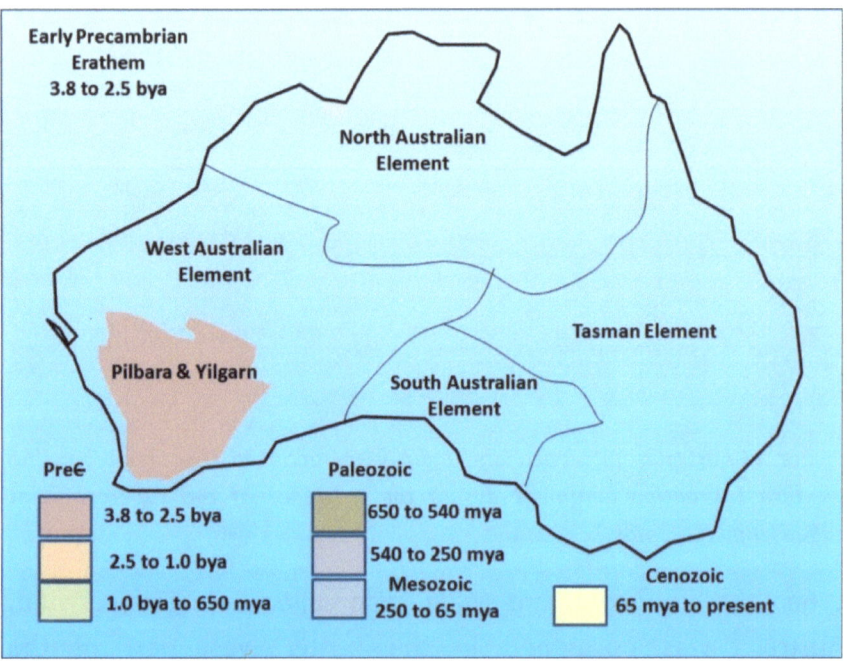

Figure 2. Four elements made up Australia. The oldest element belonged to Western Australia where the Pilbara and Yilgarn terranes accreted together first.

The Pilbara Craton developed between 3.5 to 2.9 billion years ago, overlain by 2.78 to 2.45 billion year old Fortescue and Hamersley Basins (**Figure 2**).

Origin of the craton is not clearly understood. The oldest part of the craton possibly formed by crustal overturning, or by development of an oceanic plateau resulting from hot spot plume activity, or due to tectonic processes similar to modern day plate tectonics.

Around 3.2 billion years ago, plate tectonic activity may have been active based on collisions between the Whundo greenstone oceanic arc system and Pilbara.

Pilbara was also part of the oldest known craton called Vaalbara. The Vaalbara craton included the Kaapvaal Craton of South Africa, aged 3.6 billion years. Breakup of Vaalbara began 2.8 billion years ago.

The Fortescue and Hamersley basins developed when Vaalbara broke apart. The basins represented an ancient passive margin filled with volcanic, sedimentary, and banded iron formations.

Banded iron formations were present between 2.6 and 2.45 billion years ago. Reduced iron rich bottom waters upwelled into the warmer passive margin waters, concentrating iron in reduced form. Deposition of the iron occurred between 2.6 and 1.8 billion years ago, when the Earth's hydrosphere was in an oxygen poor condition. Reduced iron was oxidized by bottom brine sediments which oxidized iron by removing silicates and carbonates from the sediment. Weathering followed upon exposure of the sediment to atmospheric conditions, which turned the iron red and removed phosphorus, leaving behind concentrated deposits of oxidized red iron.

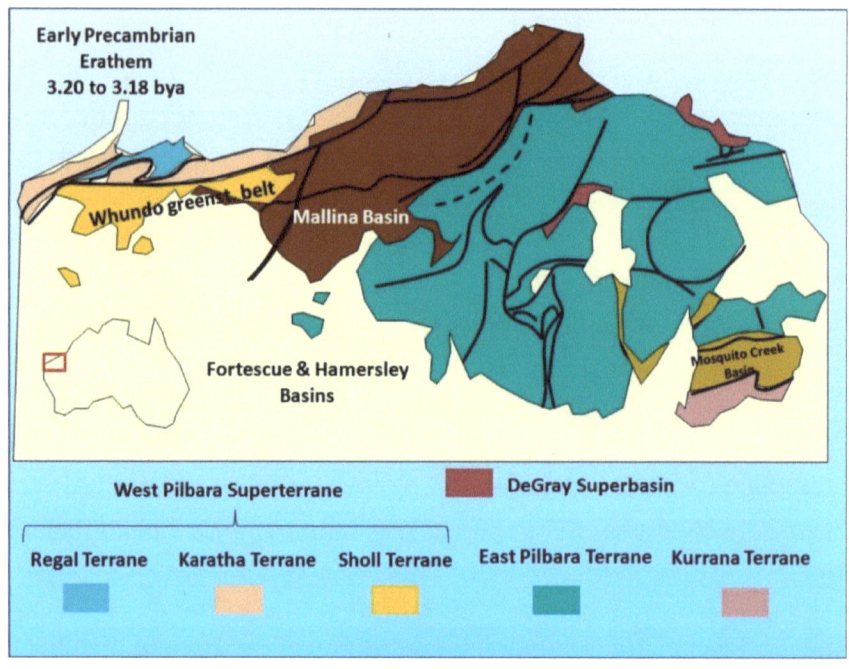

Figure 3. The assembly of West Pilbara with East Pilbara merged five smaller craton terranes together into one larger land mass. The DeGray Superbasin formed West Pilbara along the northwestern part of West Australia.

The older East Pilbara Terrane developed when granites intruded into the Pilbara Supergroup sedimentary rock column, between 3.53 and 3.17 billion years ago (**Figure 3**). A smaller granite terrane called the Kurrana terrane collided and accreted to the southeastern part of East Pilbara.

West Pilbara developed when the DeGray Super basin was intruded by granites. Smaller island arc fragments build out the western craton to the west when the Sholl, Karratha, and Royal terranes collided and accreted to the western margin of DeGray. The smaller island arcs consisted mostly of sedimentary rocks intruded by granites.

8

The Yilgarn Terrane is older in the west than in the east. The Narryer, Youanmi, and Southwest Gneiss terranes formed between 3.7 and 2.9 billion years ago.

Yilgarn is composed of several smaller terranes called the Kalgoorlie, Kurnalpi, Burtville, and Yamarna terranes belonging to the Eastern Goldfield Superterrane. The Goldfield superterrane accreted together between 2.9 and 2.6 billion years ago.

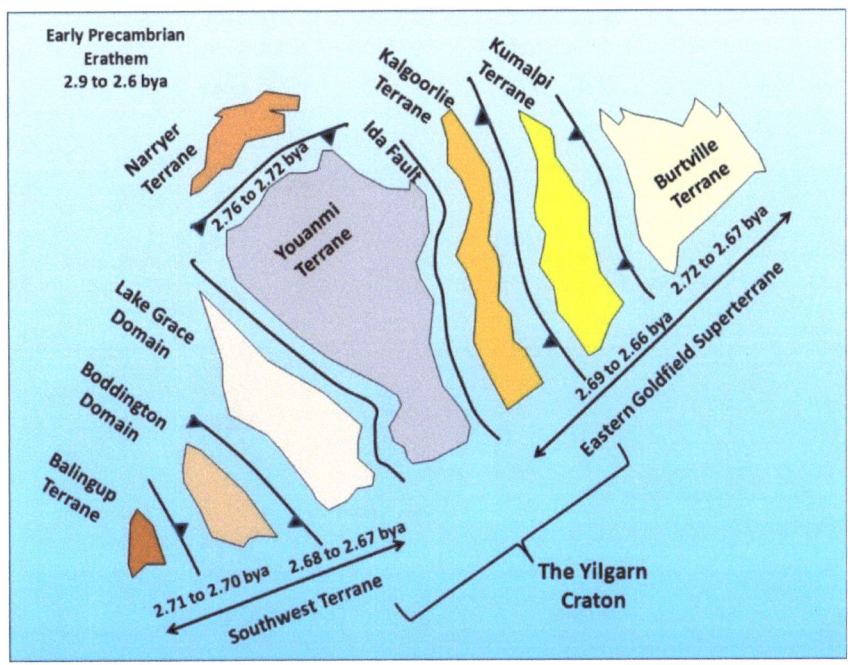

Figure 4. The Yilgarn Craton was assembled from a series of island arcs which collided by subduction and shear faulting when subduction trenches were overrun by island arcs.

Yilgarn developed during short period, catastrophic crust forming events (**Figure 4**). Final assembly occurred between 2.72 and 2.66 billion years ago. Crustal development and craton growth slowed periodically between catastrophic events.

Arc related collisions and subduction zones resulted in accretionary build out of Yilgarn. Back arc basins and major mountain building events occurred when island fragments collided together by subduction zone processes. Shear zones provided fissure eruption pathways between the mantle and land surface.

Yilgarn was part of a super-craton assemblage called Kenorland. Kenorland is thought to be a part of the Abitibi Sub-province of Canada. Kenorland accreted 2.66 billion years ago, followed by breakup 2.48 billion years ago.

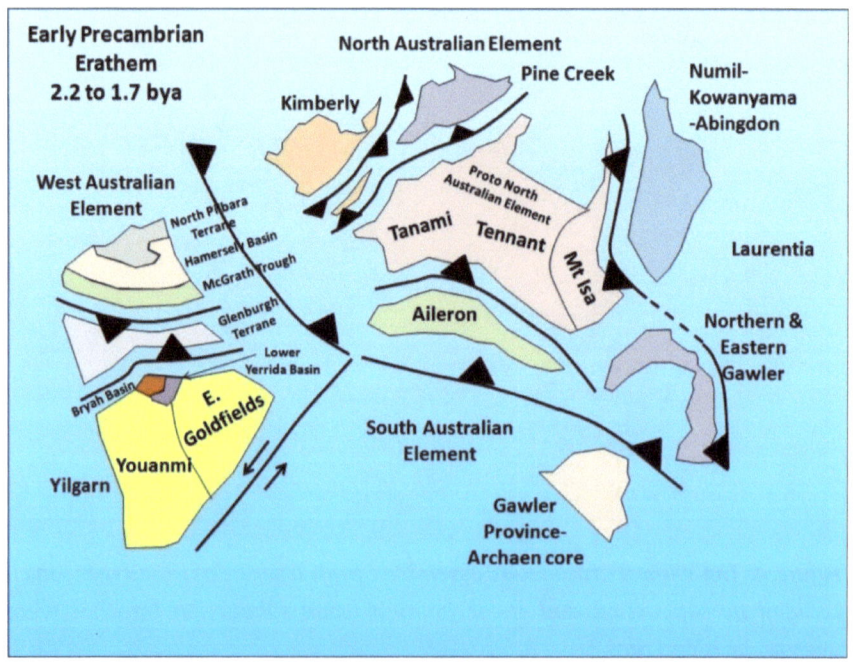

Figure 5. The assembly of Australia began when the three elements (West, East, and South Australia) collided together when Nuna accreted into a supercontinent.

Nuna (Columbia) assembled between 2.2 and 1.95 billion years ago when the three major tectonic elements collided together (**Figure 5**). The south margin of North Australia converged with Western Australia.

The Capricorn Mountains uplifted when Pilbara and Yilgarn collided, trapping the Glenburgh Terrane between the two cratons. Subduction was occurring in opposite directions beneath north Yilgarn and south Pilbara margins. The collision formed West Australia. The collision began the assembly of Nuna.

Accretion occurred between Tanami-Tennant-Isa and the Kimberly-Pine Creek provinces from the westerly direction, 1.84 billion years ago. Parallel subduction zones were driven beneath Pine Creek. At the same time, Aileron collided from the south direction. Subduction pulled Aileron to the north. Numil-Kowanyama-Abingdon collided from the east against North Australia.

A back arc basin developed along the North Australian southern margin when Aileron converged northward. Subduction from the east direction accreted the Gawler Province against South Australia. Gawler may have rifted from North Australia.

By 1.84 million years ago, West and North Australia were assembled. Between 1.81 and 1.75 billion years ago, North and South Australia accreted onto West Australia.

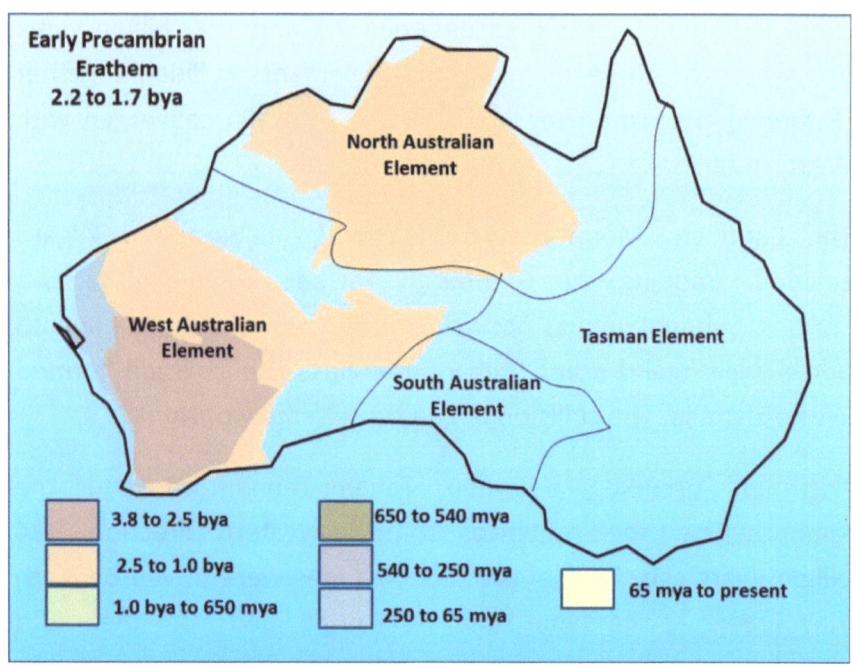

Figure 6. Assembly of Australia was nearing completion when West Australia collided with Gawler followed by convergence with North Australia.

Assembly occurred when north to northeast subduction along the south margin of North Australia allowed West Australia and Gawler to converge against North Australia. The Capricorn-Yapungku Mountains represent the sutured uplift between the two cratons.

Subduction continued beneath North Australia in the north direction after collision was completed. A left lateral strike slip fault developed along the southeastern margin of West Australia. Accretion ended when Gawler accreted onto the North and West cratonic assembly, 1.74 to 1.69 billion years ago. Final accretion uplifted the Kimban-Nimrod-Strangways Mountains.

Laurentia was attached to proto-Australia on the east up until at least 1.69 billion years ago.

12

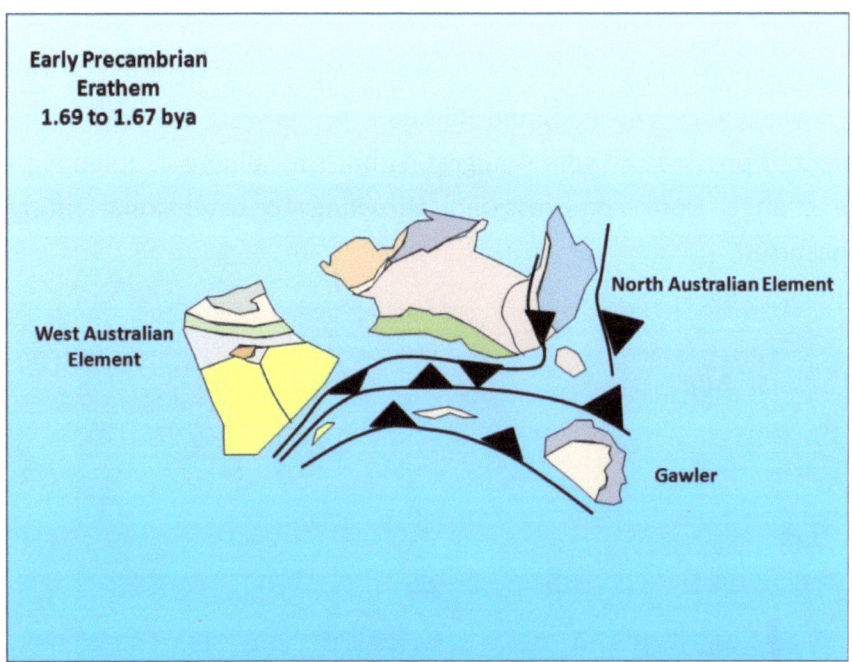

Figure 7. Gawler drifted northward on top of oceanic plates that were subducting towards the northern element. North Australia was subducting southward towards Gawler at the same time.

Nuna began to rift apart following the assembly of Australia (**Figure 7**). Laurentia rifted eastward from proto-Australia. Back arc basin development along the southern margin forced the subduction trench to jump southwards.

Aileron was intruded by granite when Laurentian rifting began. Rifting allowed seas to invade along the eastern margin of South and North Australia, developing the Calvert and Isa super-basins. Final separation of Laurentia from proto-Australia occurred to the east of Queensland.

Isa and Aileron were compressed and deformed from 1.61 to 1.59 billion years ago. Within Isa, north to south and northwest to southeast thrusting provided a mechanism for accretion of the Musgrave Province.

Gawler was exposed to high grade metamorphism at the same time when Isa and Aileron were compressed together. The Gawler Range Volcanics and Hiltaba granites were injected in the central province region. A migrating hot spot plume coupled with a change from compressional thrusting to extensional rifting occurred.

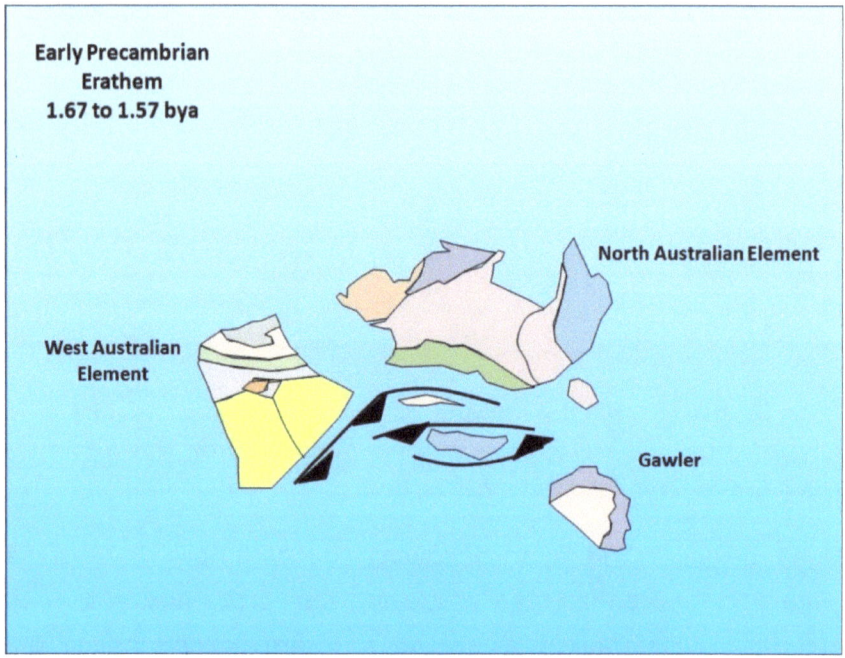

Figure 8. Gawler, Warumpi, and Musgrave continued drifting north while North and West Australia drifted southwards.

The continental back arc basin system closed when south and north directed trenches accreted rocks onto the Warumpi Province around 1.64 billion years ago (**Figure 8**). Accretions also occurred in the Musgrave and Gawler Provinces, lasting up to 1.56 billion years ago.

Gawler was part of the northward extension of the Mawson Continent. Mawson included the Terre Adelie Craton and large parts of the East Australian Shield. Mawson rifted from proto-Australia, converging again during assembly of Rodinia.

Back arc volcanism in Aileron uplifted the Chewings Mountains when Gawler converged with North Australia, 1.56 billion years ago.

From 1.64 to 1.50 billion years ago, episodic deformation of Proterozoic Australia affected North Australia. North to south directed contraction occurred along the southern margin, and in the east. The Warumpi Province accreted during this event.

Following Gawler extension, thermo-tectonic contraction occurred from east to west through the Eastern Australian basement when Laurentia converged with Australia. Hot spot movement may have been the source of thermal events.

By 1.50 billion years ago, thermal activity ended. Thermal tectonism occurs when mantle heat redistributes magma in the upper mantle by pulling up hot mantle material from below and pushing the hot material along the upper mantle until it cools and sinks back down into the mantle. The motion creates extension and contraction as the material pushes and pulls against the lower crust.

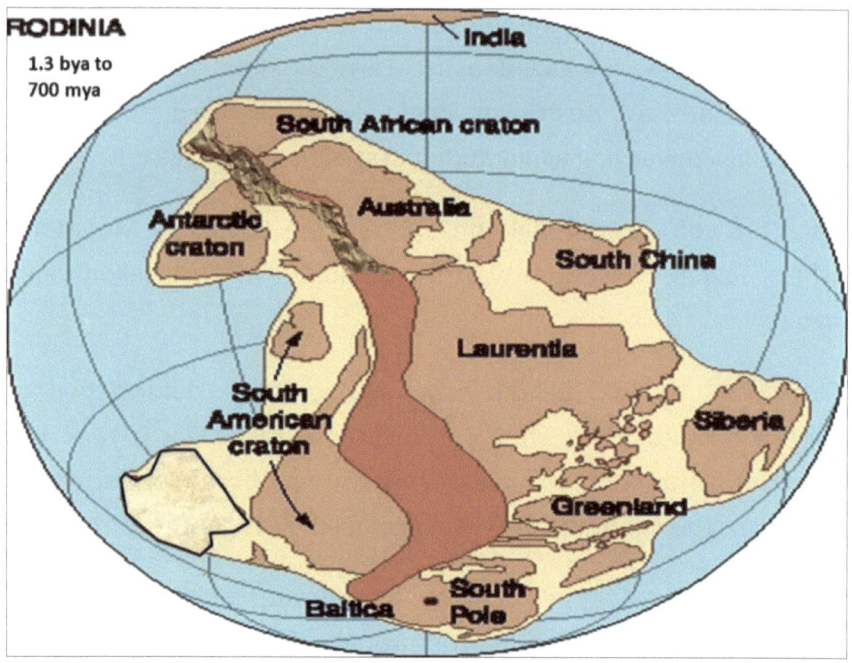

Figure 9. Rodinia assembled when the North-West Australian assembly collided and accreted onto the Mawson Craton of Antarctica. The Paterson Mountains were uplifted, part of the Grenville Uplift.

Rodinia assembled and broke apart between 1.3 billion years ago and 700 million years ago (**Figure 9**). Australia was exposed to crustal deformation which superimposed folds on top of older earlier folds from 1.35 to 1.14 billion years ago. Collision between West-North Australia against the Mawson Craton began the assembly of Rodinia. Mawson was composed of South Australia and East Antarctica. Assembly of Rodinia occurred between 1.10 billion years ago and 980 million years ago.

Rifts responsible for breaking up Rodinia developed 850 million years ago, lasting to 750 million years ago. South and North Australia were intruded by the Gairdner Large Igneous Province which extended into the Paterson Mountains in the northwest.

Central Australian Basin system development occurred by pull apart extension during rifting. The mountain chain was part of the northern extension to the Grenville Mountains.

Basin development continued into the Devonian Period in central Australia. The Officer, Amadeus, Georgina, and Yeneena basins formed along with the Adelaide Rift System. Australia was positioned adjacent to the Kalahari Craton.

The Central Australian Basin system evolved throughout the Paleozoic Era. Glacial advancement in Western Australia covered the Yeneena Basin. Both the Adelaide Rift System in South Australia and King Island in Tasmania were impacted.

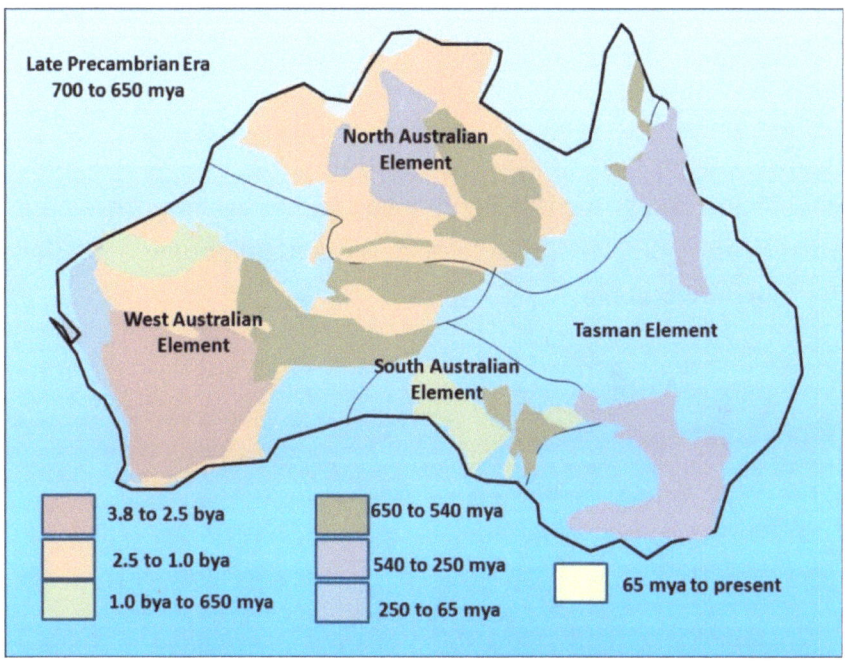

Figure 10. The Paterson Province and South Australia (pale green & pale brown) were intruded during assembly of Gondwana, 650 million years ago. Petermann Uplift (lavender) followed in North Australia. A mantle hot spot plume is suspected of being the source of granitic composition.

West Australia was deformed between 650 and 600 million years ago when granites were intruded into the Paterson Province (**Figure 10**).

Folding and faulting intensified when the King Leopold and Petermann Mountains uplifted in the Kimberly and Musgrave Provinces, 560 to 525 million years ago. Uplift occurred from the northwest into central Australia. Uplift of the Petermann Mountains was possibly related to the Kuunga Mountain uplift during collision between Australia and India when Gondwana assembled.

Further evidence of Gondwana assemblage occurred along the southern part of the Australian west coast. Uplift of the Punjarra Mountains contains elements of collision zones between Australia-Antarctica and India-Antarctica.

Following uplift of the Punjarra Mountains, 510 million years ago, igneous intrusions were injected into the Kalkarendji province in western and north central Australia. A mantle plume was possibly the source of magma.

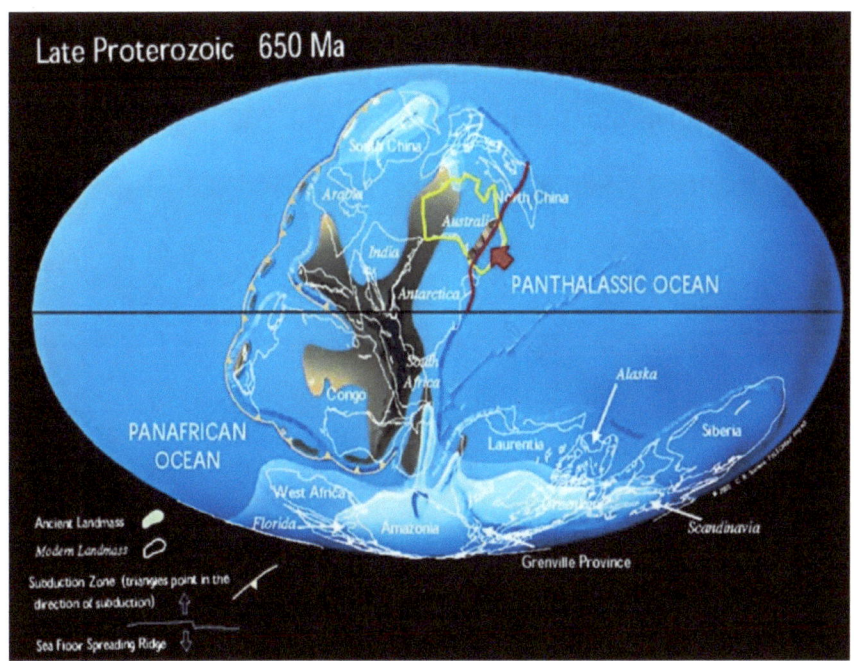

Figure 11. The Panthalassic trench intersected with submerged Eastern Australia, uplifting the Tasman Element of Australia. Arc magmatism and back arc extension developed above a westward subducting oceanic plate.

The former passive margin of Eastern Australia became an active mountain belt called the Terra Australia Orogen, covering the entire Gondwana margin (**Figure 11**). Several cycles produced the Tasman Element, beginning with arc magmatism and back arc extension related to west dipping subduction. Continental slivers and island arcs accreted back onto Australia, uplifting and thrusting mountains landward.

Five cycles were recognized: Delamerian (515 to 490 million years ago); Benambran (490 to 440 million years ago); Tabberabberan (440 to 380 million years ago); Kanimblan (380 to 350 million years ago); and Hunter Bowen (350 to 220 million years ago). Some of the younger cycles were superimposed on top of older cycles.

Cycles resulted from long lived subduction retreating, associated with extension cycles interrupted by short lived subduction advances involved contraction, or thrust faulting.

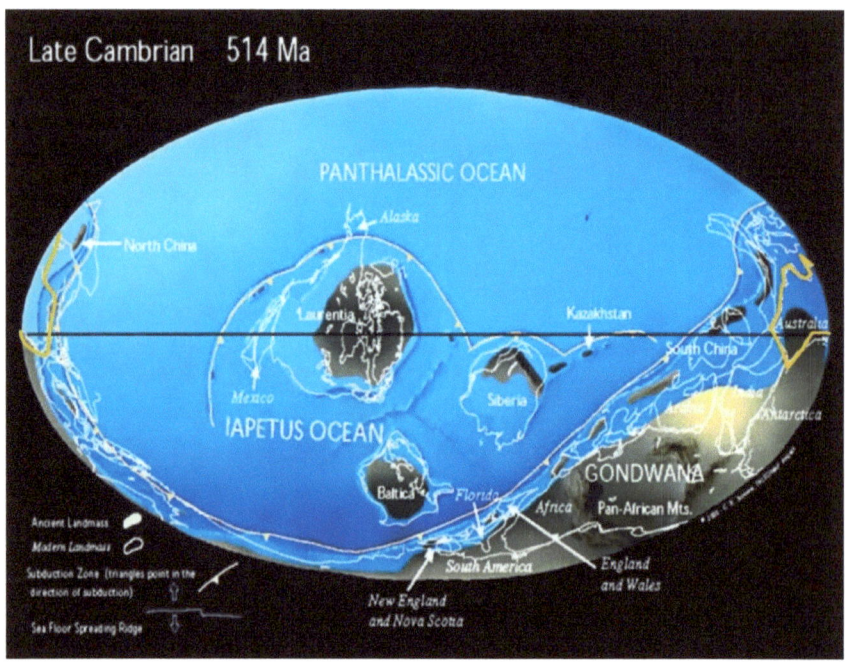

Figure 12. Australia was positioned north of the Equator, attached to the northern end of Gondwana.

During the Late Cambrian Period, most of Australia was submerged beneath a shallow sea. Western Australia and portions of South Australia were emergent. Western Australia was an extension of Antarctica, positioned at the Equator.

Antarctica was emergent at this time, positioned in the high southern latitudes beyond the reach of South Polar ice. India was attached to northwestern Antarctica, completing the composition of Eastern Gondwana.

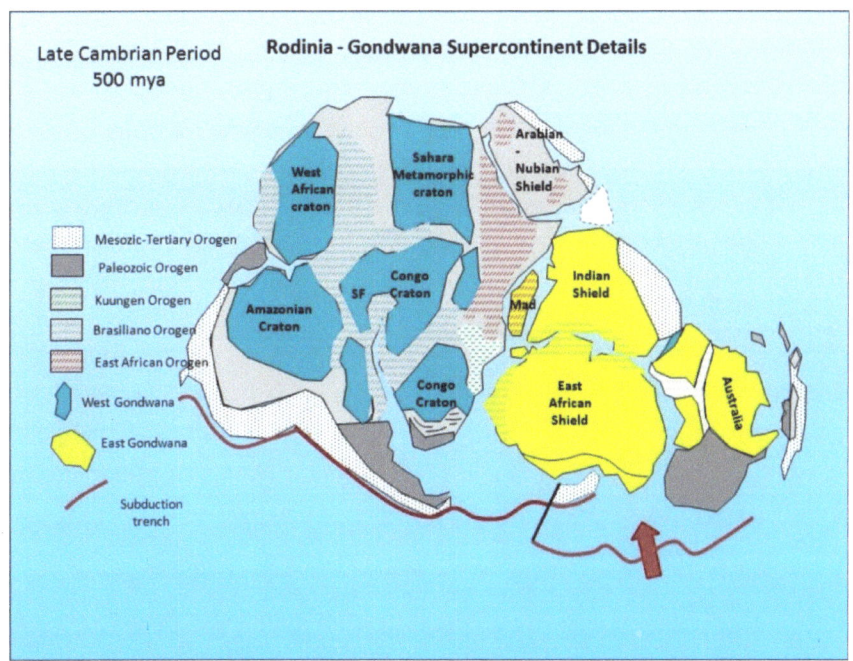

Figure 13. By the end of the Rodinian supercontinent, Gondwana was divided into an eastern and western part. Australia-Antarctica was attached to East Africa. India was attached to the northern part of East Africa.

By the end of the Cambrian Period, 500 million years ago, Australia was part of East Gondwana, accreted to India and West Africa (**Figure 13**). Island arcs and small continental fragments remained off the northeastern Australia margin.

Tasman built out Eastern Australia. The central subduction zone jumped to an offshore position away from the eastern margin, continuing to drive oceanic crust beneath the shelf.

Back arc basins formed during early cycles coupled with later contraction deformation. Contraction deformation is another term used to describe folding related to thrust faulting.

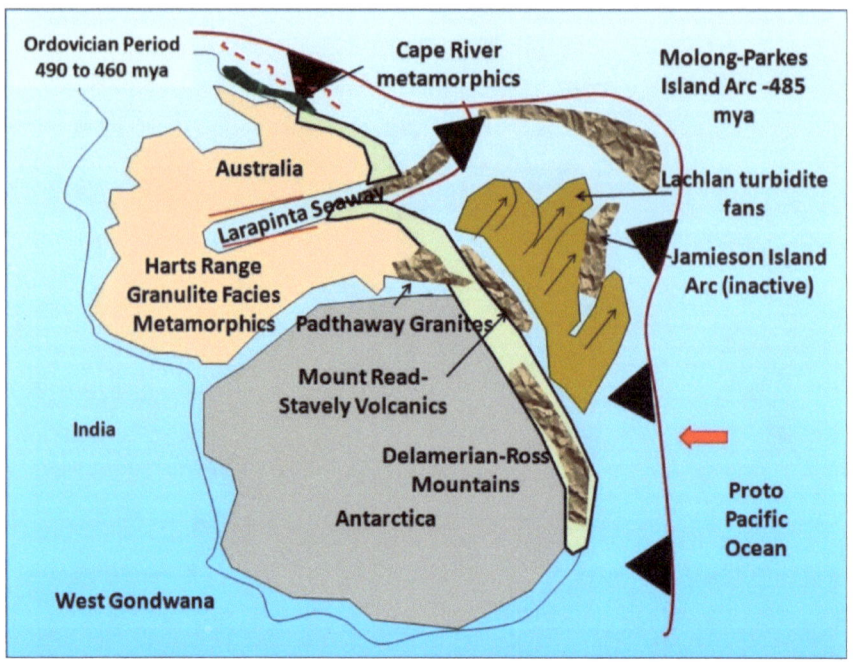

Figure 14. Westward subduction of the Proto Pacific plate against Australia, and northward subduction along the Eastern Australia continental shelf created a rift through the central interior of Australia, allowing the Larapinta Sea to invade into central Australia.

Central Australia was exposed to interior plate deformation during the Paleozoic Era (**Figure 14**). Convergence along the eastern margin impacted the Australian interior.

The Larapinta Seaway and the Alice Springs Mountains were the two significant events that occurred between 490 and 460 million years ago. Ordovican marine sediments were deposited in the sea connecting Georgina and Amadeus basins to Eastern Australia. Seas were warm and shallow, rich in organic deposits.

Larapinta sediments were metamorphosed into granulites during a rifting event in the Irindina Province. A deep sub-basin occupied the province within the Larapinta Seaway.

22

Basaltic eruptions occurred early in the sea during a rifting event which invaded the Antrim Plateau Volcanics in the Kalkarindji Large Igneous Province.

Australia was attached to Antarctica during the Ordovician Period. Small island arcs were colliding with Australia-Antarctica by westward subduction of the Proto Pacific Plate towards West Gondwana. Large sedimentary wedges eroded from the shelf area, cascading down into the deeper back arc basin area, developing turbidite deposits.

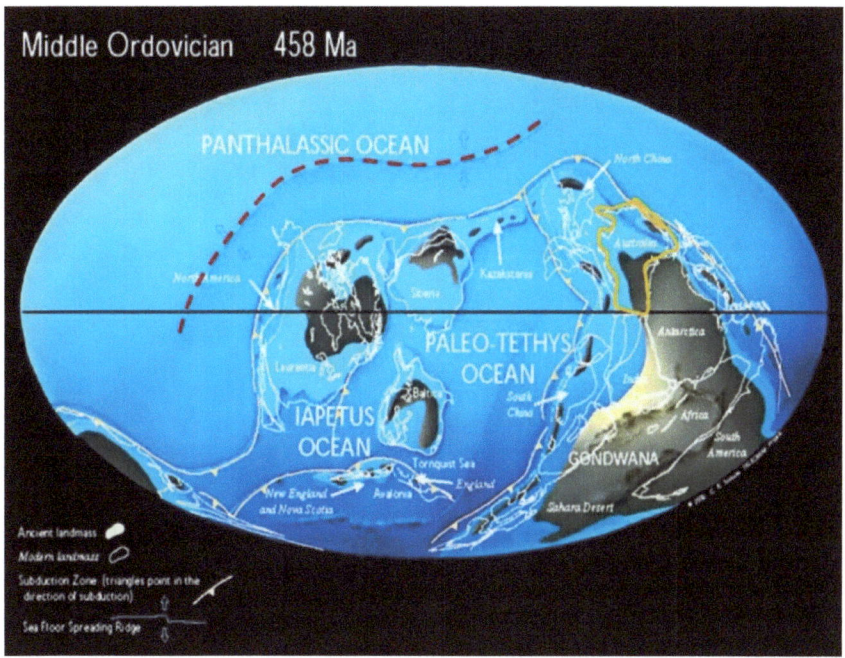

Figure 15. Australia was pushed northward above the Equator by the rotation of Gondwana. Seas invaded most of the northern and central parts except for Southwest and South Australia.

Gondwana was rotating counterclockwise, carrying Australia-Antarctica along the leading edge of the super continent (**Figure 15**).

Seas invaded Western Australia, covering portions of northern South Australia. Australia was pushed further north of the Equator.

The western Panthalassic Ocean trench was positioned off the eastern submerged margin. A new spreading ridge opened up in the southern part of the Panthalassic Ocean, pushing southward against Laurentia, Siberia, and Gondwana.

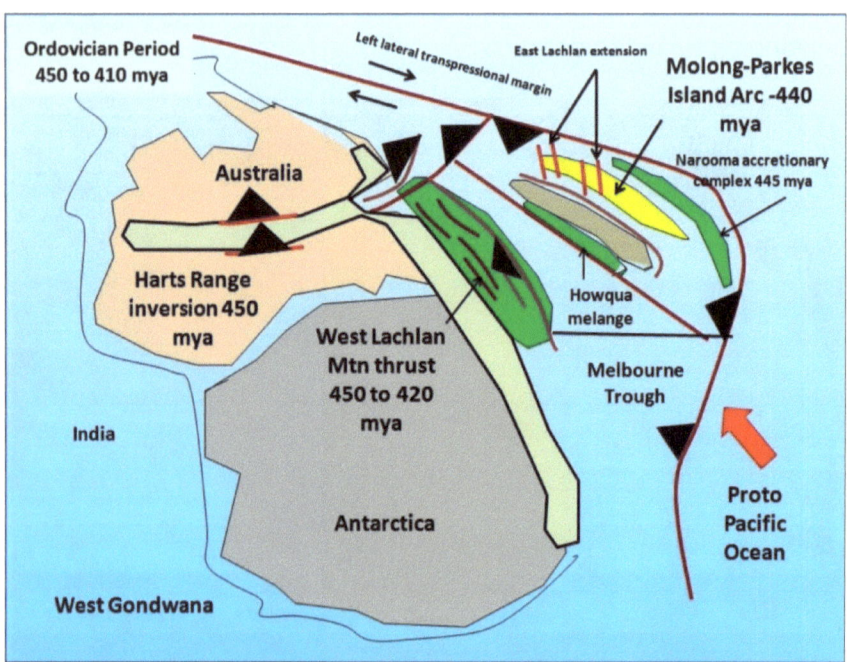

Figure 16. The Larapinta Sea dried out and became wedged between parallel thrust belts in central Australia, inverting the Harts Range. Central thrusting uplifted the West Lachlan Mountains along the East Australian shelf.

Translational compression along the northeastern margin of Australia began to deform the former Larapinta Sea sedimentary deposits (**Figure 16**). Sediments were uplifted and thrusted to the south, inverting the Harts Range, 450 million years ago.

Larapinta thrusting forced the West Lachlan Mountains to compress and deform into a mountain belt adjacent to the margin shelf.

A smaller subduction trench was positioned along the eastern edge of the West Lachlan Mountains, between 450 and 420 million years ago.

Subduction of the Proto Pacific Plate changed direction from west to northwest. The Molong-Parkes, Narooma and Howqua island arcs began to accrete together off the eastern Australian-Antarctica assembly, 445 to 440 million years ago. This event was called the Rodingan Movement.

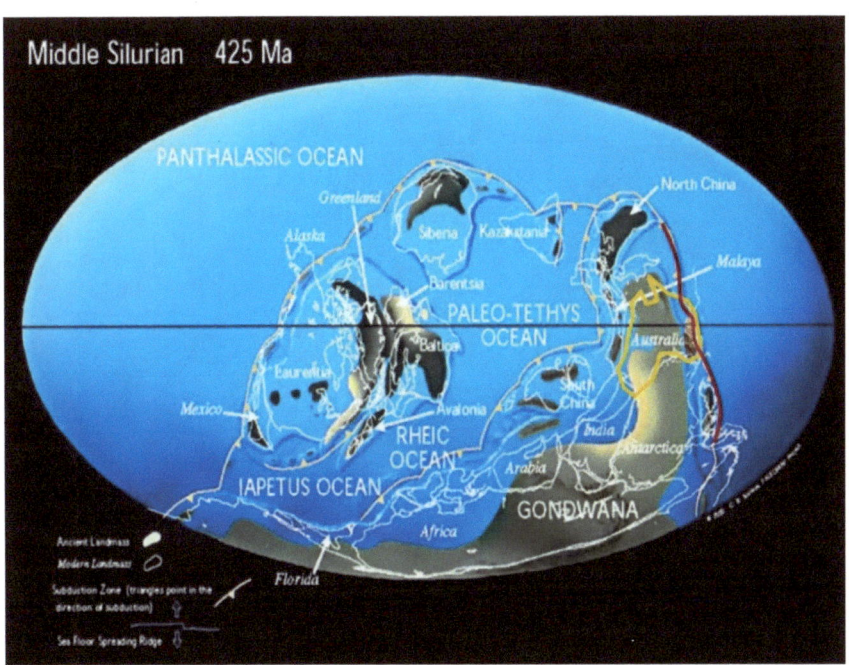

Figure 17. Australia was positioned at the Equator. The Panthalassic trench intersected the eastern margin back arc basin, submerged beneath a shallow sea.

The Panthalassic spreading center was extinguished by the Siberian trench portion of the subduction zone when it overrode the spreading center (**Figure 17**).

Gondwana reversed direction, migrating southward in a clockwise rotation from subduction occurring from the Siberian trench and North China portions of the trench.

Seas withdrew from most of interior Australia except for East Australia, and a smaller embayment which continued to submerge a small region of West Australia.

The west Panthalassic subduction zone was positioned beneath submerged East Australia. An island arc developed above the trench. The back arc region to the island arc remained submerged.

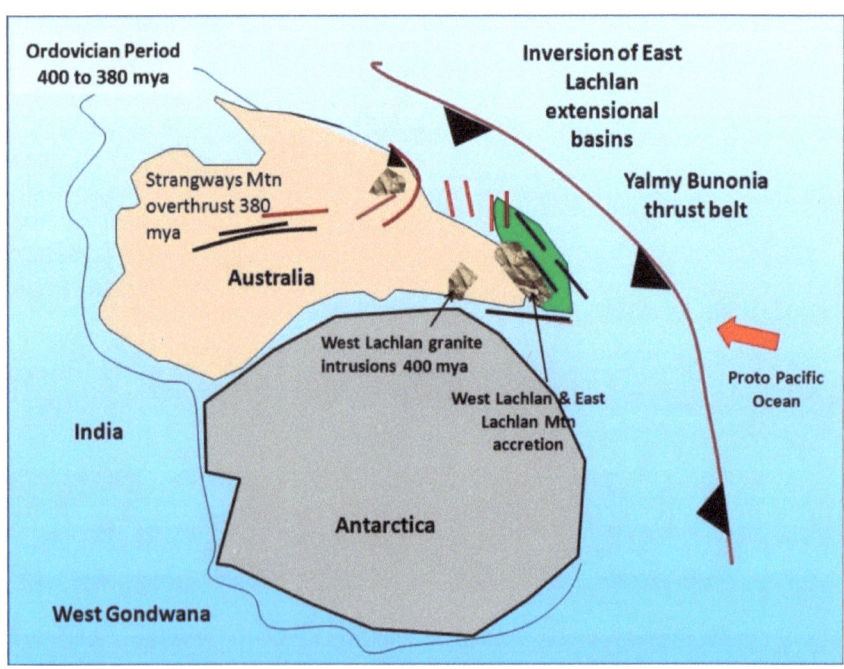

Figure 18. Thrusting accompanied by westward subduction continued along the eastern Australian margin, uplifting mountains, injecting granites, and creating small extensional basins.

The Pertinjara Movement (390 to 380 million years ago) was the second compression event responsible for uplifting the Alice Mountains (**Figure 18**). Deep crustal rocks in the Irindina Province were brought to the surface in the Georgina Basin.

Consequently, the Redbank Shear Zone developed, offsetting the Moho discontinuity, squeezing upper mantle material into a small thrust belt positioned along the Eastern Australian margin.

Proto Pacific Ocean plate subduction continued to the northwest, beneath the Australia-Antarctica shelf margin. The Yalmy Bunonia thrust belt formed when the island arc collided against southeastern Australia. The collision forced West Lachlan into the Central Lachlan Mountains, injecting granites above the down going oceanic slab beneath the margin.

The East Lachlan Mountains collapsed from extension occurring above the subducting trench beneath northeast Australia. Further inland, the Strangways Mountains were over thrusted to the north.

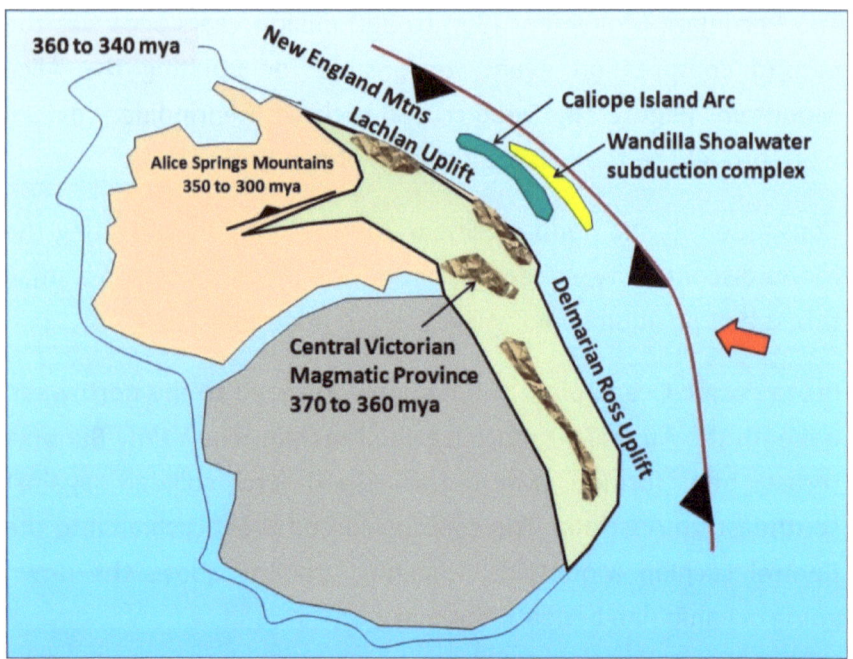

Figure 19. The New England Mountains uplifted during the last movement event, called the Brewer Movement.

The last movement event was called the Brewer Movement (**Figure 19**). Uplift and erosion created a foreland basin ahead of a continental arc system, resulting in the development of the New England Mountains. The foreland basin subsided when counter flow in the upper mantle above the subduction zone deflated the overlying crust.

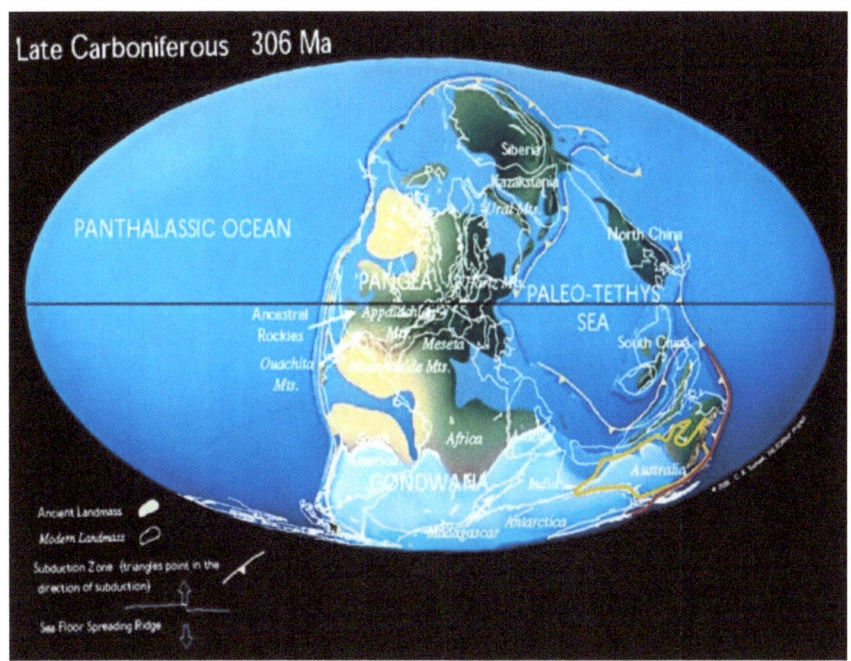

Figure 20. Glacial ice advanced across most of Australia except for a small portion of North Australia.

Rotation continued (**Figure 20**). Australia-Antarctica continued to drift into the South Polar region. The island arc positioned above the eastern subduction zone collided with East Australia. Australia was emergent. Glaciers advanced across Western, and Southern parts of Eastern Australia, and parts of Northern Australia. Antarctica and India were entirely covered by glacial ice, including parts of Africa.

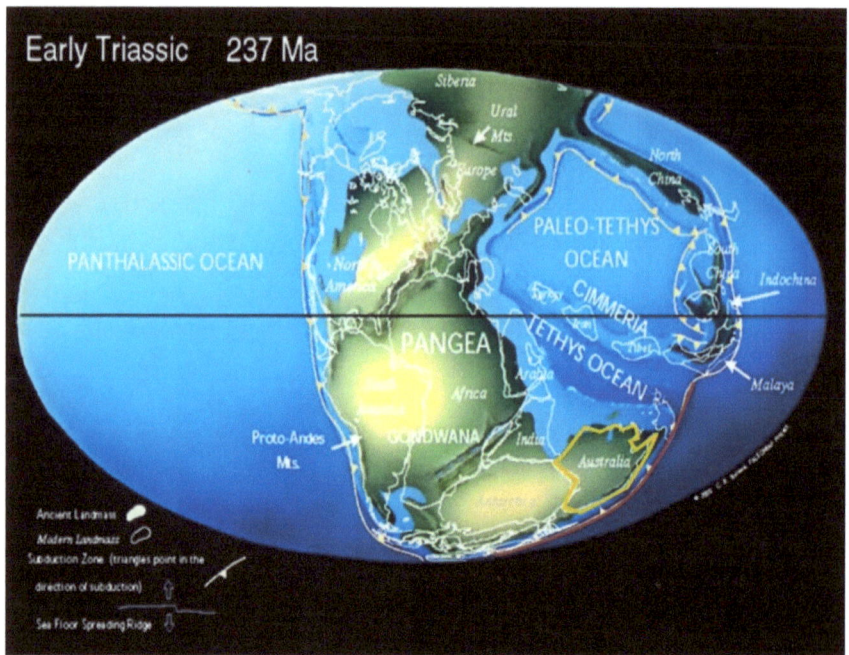

Figure 21. Rotation of Gondwana in a counterclockwise direction pushed Australia out of range from the glacial ice at the South Pole. Subduction of the Panthalassic trench continued beneath Eastern Australia.

Australia began to drift in a more westerly direction while attached to Gondwana. Glaciers retreated from most of the South Polar region, leaving the region emergent above sea level. The shift in position allowed a back arc basin to form between the Panthalassic trench and the East Australian margin. Subduction occurred beneath the Australian eastern margin. Pangea assembled.

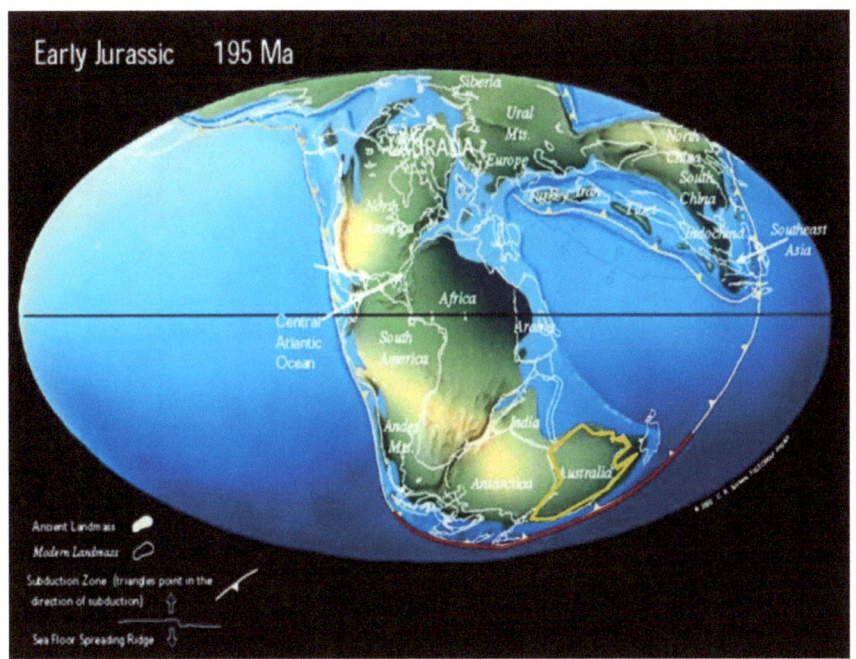

Figure 22. Rotation of Gondwana continued pushing Australia northward. The breakup of Gondwana began.

Pangea remained stable between 250 and 150 million years ago (**Figure 22**). Around 180 million years ago, Gondwanan break up began when the Atlantic Ocean and Tethys Sea rifted open, separating Gondwana.

Australia, South America, Africa, India, and Antarctica split from Europe and North America. Rifting began along the eastern African coastline, beginning the separation of Australia- India and Australia-Antarctica from Pangea.

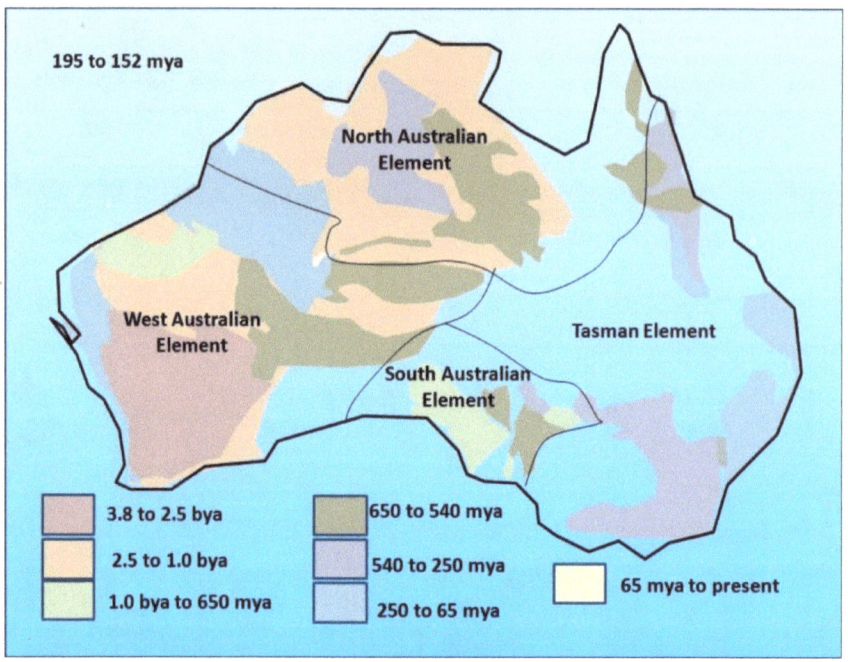

Figure 23. Intrusions accompanied rifting of Gondwana. Tasmania was part of the rifting process when basaltic dikes and sills were injected into the crust (light purple).

Tasmania was intruded by large volumes of basaltic rock called dolerite (also called diabase) during the rifting of East Gondwana from West Gondwana (**Figure 23**). The Karoo Ferrar Large Igneous Province was formed by these injections.

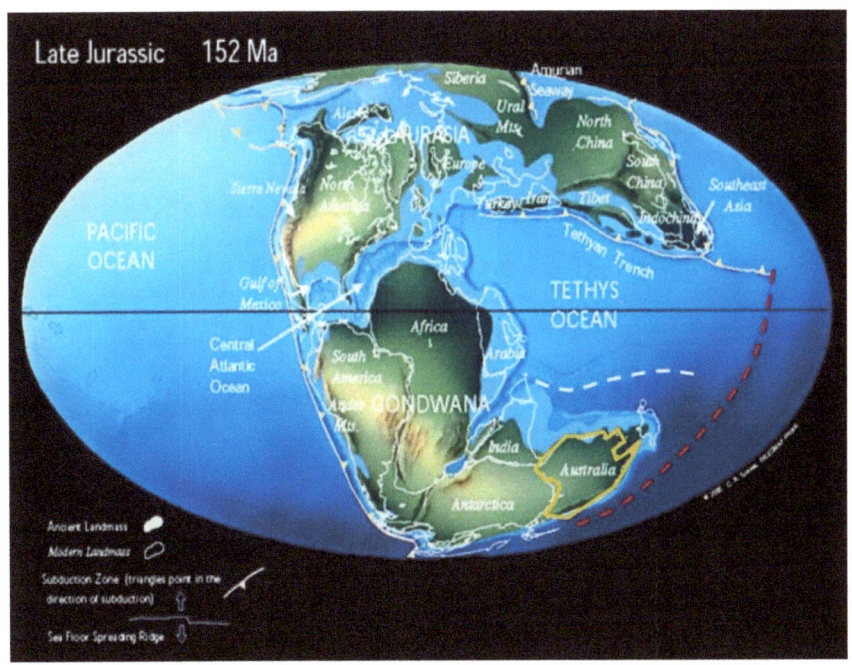

Figure 24. Gondwana continued rotating to the north, pushing Australia along the leading edge. Antarctica-Australia began to separate from Africa. A new spreading ridge developed in the southern Tethys Ocean.

The former Pacific subduction trench converted into a spreading center in the western Pacific Ocean (**Figure 24**). The ridge system was positioned off the eastern Australian margin. Antarctica, India, and Australia remained attached. The Pacific spreading center provided the mechanism for clockwise rotation of Australia-India-Antarctica. India and Antarctica started to rift away from Eastern Africa. A new spreading ridge began to develop in the southeastern Tethys Ocean off the northwest Australian coast.

India separated from Western Australia along a triple extension junction which included the separation of Antarctica from Africa. The northwestern Australian margin developed when the last of a series of continental slivers rifted away to form Argo Land in Burma, 155 million years ago.

Australia separated from Antarctica along an easterly expanding rift. Between 155 and 145 million years ago, rift basins expanded into South Australia and in the Victoria region.

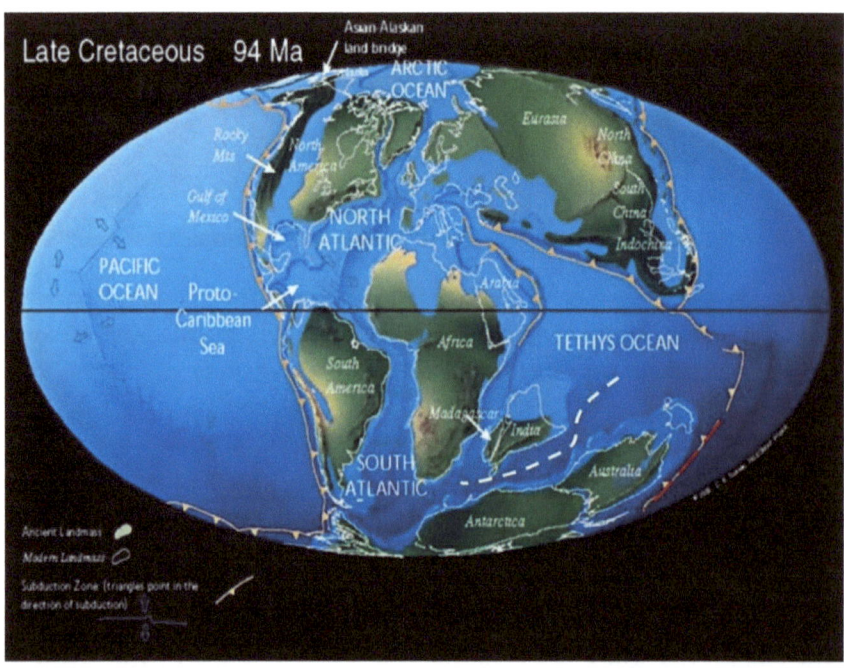

Figure 25. The Indian Ocean spreading center opened between India and Australia-Antarctica, pushing the land mass east while Pacific trench subduction was driven beneath East Australia.

Australia and Antarctica rifted from India along a new spreading ridge center which opened the Indian Ocean. India separated from Africa (**Figure 25**). Australia-Antarctica drifted to the east.

The Pacific spreading center fragmented into short subduction trench segments off the eastern Australian coast. A shallow sea invaded the northeastern part of Australia.

The breakup of Gondwana from the North West Shelf produced rifting along the Northwest-Southeast Westralian trend. Deep marine basins collected organic rich sediments.

Organic sediments were buried by the Early Cretaceous Barrow Delta when seas retreated. Following continental breakup, Cretaceous marine shales were deposited on top of the organic deposits. The separation of Greater India from Australia was completed by 120 million years ago. The Kerguelen Large Igneous Province erupted beneath the Southern Ocean, 110 million years ago.

Central and Eastern Australia were submerged beneath a large inland sea. The Eromanga and related basins developed in response to corner flow in the asthenosphere beneath the eastern Australian crust during west directed subduction.

By Late Cretaceous time, 95 million years ago, subduction jumped eastward ending corner flow while triggering uplift of the Eastern Australian highlands. The Eromanga Basin became inverted by the uplift. Erosion of the basin shed sediments into a large deltaic system in the offshore region of Ceduna, South Australia.

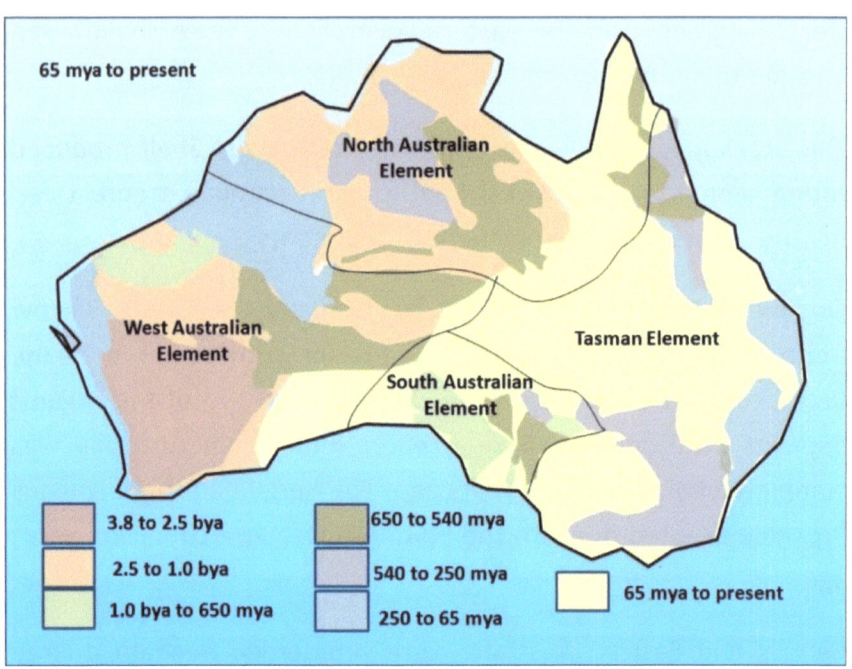

Figure 26. The Tasmanian Element was completed 65 million years ago when Australia separated from Antarctica. Wind, lake, and river deposition contributed to Cenozoic Era sedimentation across Antarctica.

A large seaway developed between Australia and Antarctica. Tasmania remained connected to Antarctica (**Figure 26**). Rifting between Australia and Antarctica separated the two continents along the Tasmania-Victoria rift. The Otway, Bass, and Gippsland basins formed by collapse along the Tasmania-Victoria Rift.

Seafloor spreading along the east coast began 84 million years ago in the south, expanding northward. The Tasman Sea opened and rifting began along the Lord Howe Rise and in New Zealand. Rifting ended 56 million years ago. The Coral Sea rift began to separate further north at the same time the Queensland Plateau rifted from the northeast coast of Queensland.

Weathering in Australia began when Gondwana broke apart. Clays were deposited during the Mesozoic in the lower latitudes of Australia. By the beginning of the Cenozoic Era, 65 million years ago, Australia was covered by thin aeolian (wind), lacustrine (lake), and fluvial (river) deposits. Many basins developed including the Eucla, Murray, Lake Eyre, and Karumba basins.

Seafloor spreading in the Southern Ocean accelerated 45 million years ago. Full separation between Australia and Antarctica was completed by 34 million years ago.

Between 45 and 10 million years ago, deposition in the Lake Eyre, Eucla, Karumba, and Murray basins was completed. Climate was warmer and wetter with extensive rain forest coverage.

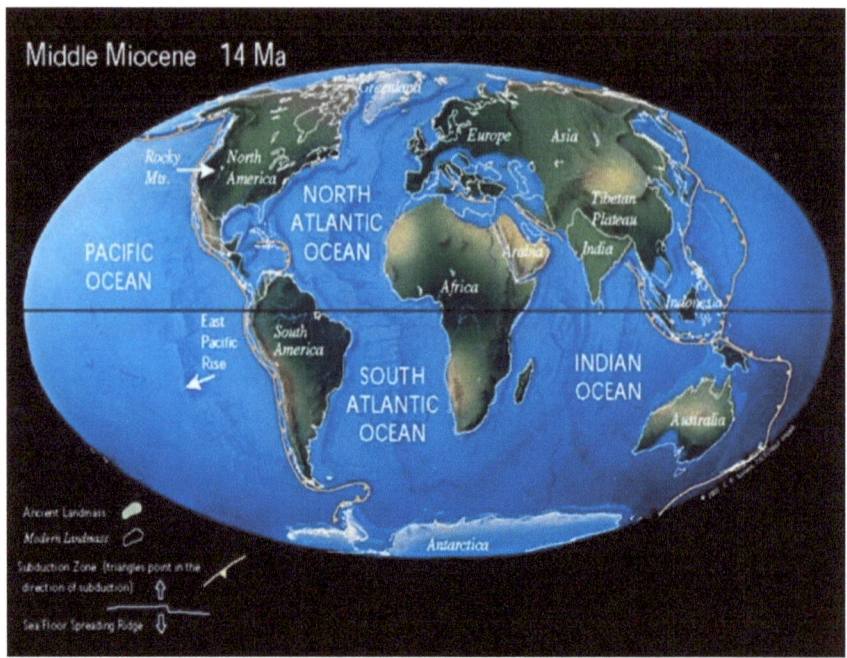

Figure 27. Australia was a separate continent by the Middle Miocene Epoch. Erosion of the continental interior transported and deposited sediments internally during humid climate periods. Short term sea advance and retreat cycles covered the continent by Late Miocene time, covered over by glaciers during the Holocene Epoch.

Between 30 and 20 million years ago, river channels drained the continental interior leaving behind buried paleo-channels and floodplains rich in iron, organics, and heavy minerals in the Pilbara Craton (**Figure 27**). Organic rich sea deposits formed in the La Trobe Valley in the southeastern Victoria swamps when the climate was humid. Lacustrine basins were exposed to rifting near Gladstone and Prosperpine in Queensland. Shales accumulated.

By 25 million years ago, Australia and Antarctica separated enough allowing full circulation to occur in the Southern and Pacific Oceans, affecting climates of both continents.

From 15 to 10 million years ago, the great ice sheets of Antarctica developed in the east. The western ice sheets formed between 10 and 6 million years ago.

Australian climates changed drastically from Antarctic glaciation. Subtropical monsoons shifted northwards and the continent dried out. The middle latitudes became arid in Australia, Africa, North and South America. Rainforests disappeared.

During the Eocene and Miocene Epochs, 15 to 5 million years ago, seas advanced overland several times. Heavy mineral marine sediments mixed in with continental sediments.

Tasman was exposed to large scale volcanism. A series of hot spot tracks coupled with deep mantle upwelling marked the northward migration of Australia.

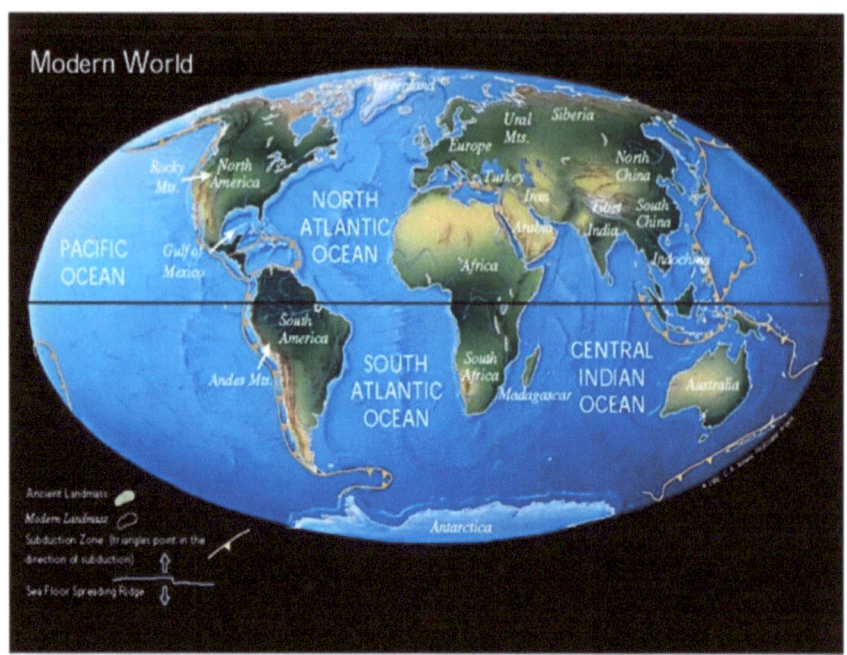

Figure 28. Australia is drifting to the north against the Indonesian plate, uplifting New Guinea and other island arc systems above the Indonesian trench.

Presently, convergence occurs along the northern Australia margin. Timor is part of the Australian Plate. New Guinea uplifted from the colliding Indonesia island arc against the Irian ophiolite complex, forming the mountainous spine of the Guinea island arc system. Southern Guinea became an active fold thrust belt, developing active foreland basins.

Left lateral faulting developed across the Australian Plate boundary in New Zealand. Crustal shortening in Australia resulted in uplift of the Flinders Ranges.

Tasmania was exposed to small glacial advance cycles. The maximum advancement occurred 1.0 million years ago. Later smaller advances peaked 100,000 years ago.

Arid climates occupied the interglacial periods. Vast areas of sand dunes developed across Australia.

Chapter II: Antarctica

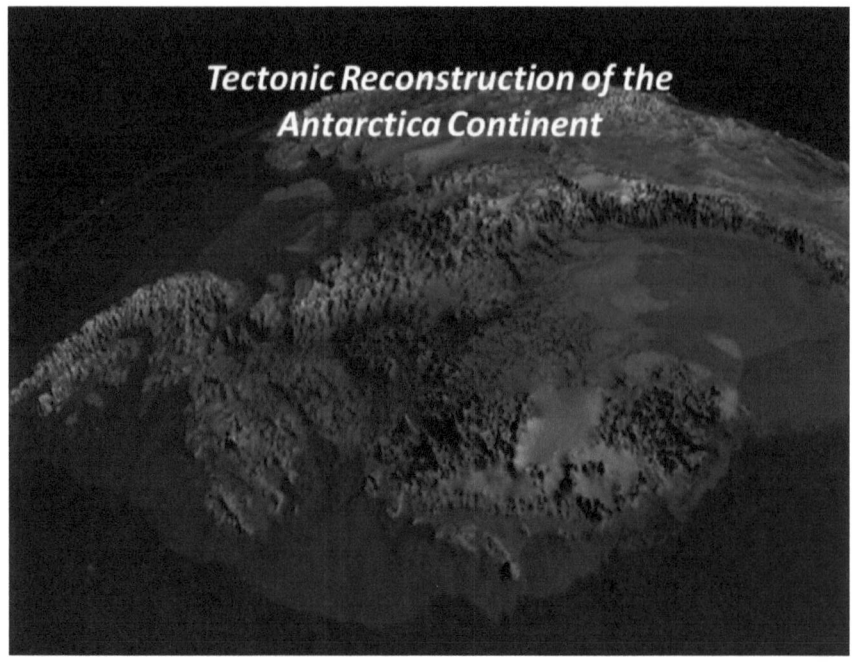

Tectonic Reconstruction of the Antarctica Continent

Precambrian Era history is masked by multiple episodes of metamorphism in Eastern Antarctica but matches with Western Australia, South Africa and associated terranes of Gondwana.

Sedimentary and igneous rocks were exposed to high temperatures and pressures, altering their original features. Evolution spans a period of time beginning 3.1 billion years ago to 480 million years ago.

Around 550 million years ago, near the end of the Precambrian Era into the Early Paleozoic Era, marine sediments and volcanic rocks suggest an active plate margin was present. Mountain building events deformed and metamorphosed rocks.

By Devonian time (400 million years ago) East Antarctica emerged above sea level as a stable continent within interior Gondwana. Precambrian mountains were eroded down to flat plains called peneplains.

From the Devonian into the Jurassic Period (380 million years ago) flat lying sedimentary rocks were deposited on top of the peneplain. Swamps occupied the Transantarctic and Prince Charles mountain regions. Rivers and floodplains dominated the region.

During the Permian Period (290 million years ago) glaciers advanced across Antarctica. Glaciers covered most of Gondwana when Gondwana drifted into the South Polar region.

In Early Jurassic time (180 million years ago) Gondwana began to break apart. Magmas were intruded and extruded on top of the worn down Transantarctic Mountain region. The Dufik Massif was one of the regions intruded, forming the Pensacola Mountains in Coats Land. Sills and dikes also intruded into Victoria Land.

Gondwana split apart, drifting north away from Antarctica. Antarctica remained in the polar region. South Africa separated from Antarctica 145 million years ago followed by India, Australia, and South America, ending 29 million years ago.

Marine sediments mixed with volcanics in the Antarctica Peninsula during the Mesozoic Era. Granites were intruded into the Peninsula, metamorphosed during the Antarctica Andean Mountain Uplift, named for similarities with the South American Andean Mountain tectonism. Marine sediments accumulated east of the Peninsula in island and offshore basins. Continental sediments mixed in with marine deposits.

The West Antarctica Rift System and plate collisions in the Peninsula region were exposed to volcanic activity. The rift system remained active for about 50 million years. Uplift along the western flanks of the rift produced the Transantarctic Mountains. Much of the range is buried under the East Antarctica Ice Sheet.

Basalt volcanoes appeared in the McMurdo Sound and Victoria Land regions. In Marie Byrd Land, at the opposite end of the rift system, volcanoes developed beneath the ice sheet.

At the start of the Cenozoic Era (65 million years ago) global cooling began, interrupted by warm interglacial periods.

Large scale glaciation began 35 million years ago. Cool temperate forests were present. Tundra vegetation replaced forests when the climate continued to cool. Glaciation cycles were common up until 15 million years ago when the climate was cold. Glacial ice expanded to its maximum size, burying all mountains ranges, covering the continental shelf areas.

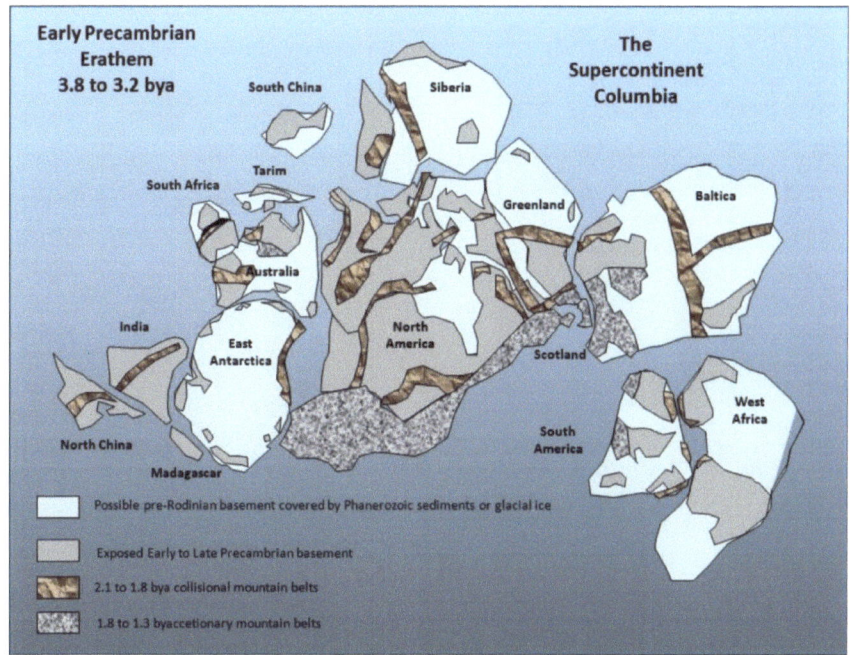

Figure 1. East Antarctica occupied the southern part of the Australia-Antarctica continental land mass. India was attached to West Antarctica. Laurentia was positioned along the eastern margin.

During the Early Precambrian Era, East Antarctica was assembled with Australia along the western margin of Laurentia (**Figure 1**). India was attached to Antarctica. Antarctica was oriented in a north-south position.

The margins were occupied by small cratons consisting of Early to Late Precambrian basement rocks rimming the north, west, and south margins of the main continent. Between 2.1 and 1.8 billion years ago, a mountain belt uplifted along the eastern Antarctica margin following collisions with Laurentia. Cratons, positioned along the northern margin, probably had a close affinity to cratons located along South and West Australia.

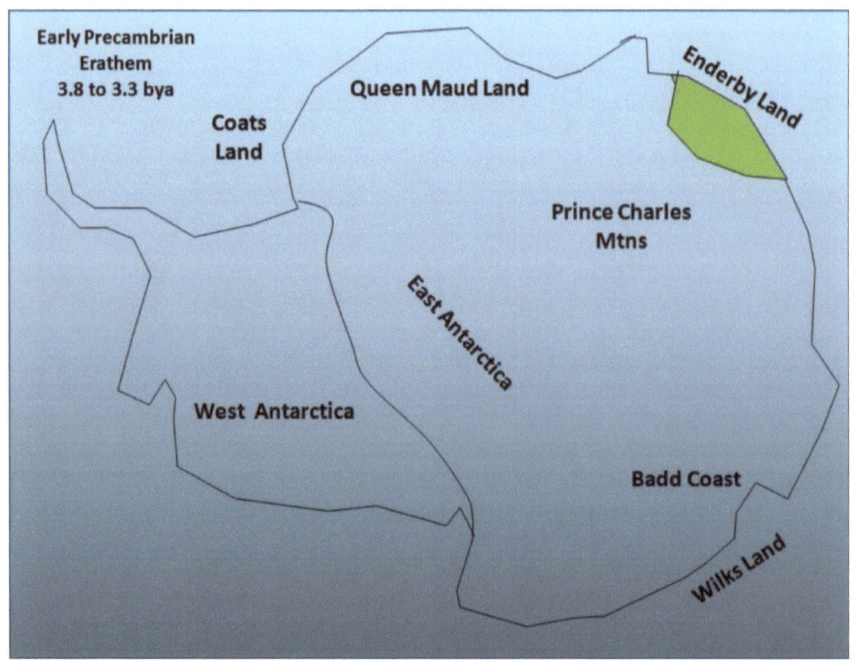

Figure 2. Enderby Land was the first craton to develop the Antarctica continent. The Napier Complex is the oldest of the cratonic terranes within Enderby Land.

The Napier Complex in Enderby Land was deformed and metamorphosed by high intensity heat and pressure, 3.8 billion years ago. Metamorphism occurred again between 2.98 and 2.85 billion years ago. Flat lying and isoclinal folds were subjected to multiple refolding and reorientation, destroying original structural signatures. Metamorphism occurred between 24 and 40 km within the crust, followed by long term cooling at considerable depths (20 to 30 km), ending in the Middle Precambrian. Dike swarm intrusion followed during the Grenville Mountain uplift, 1.0 billion years ago. Burial depths suggest metamorphism occurred deep within a subduction trench or within the continental crust influenced by plate subduction.

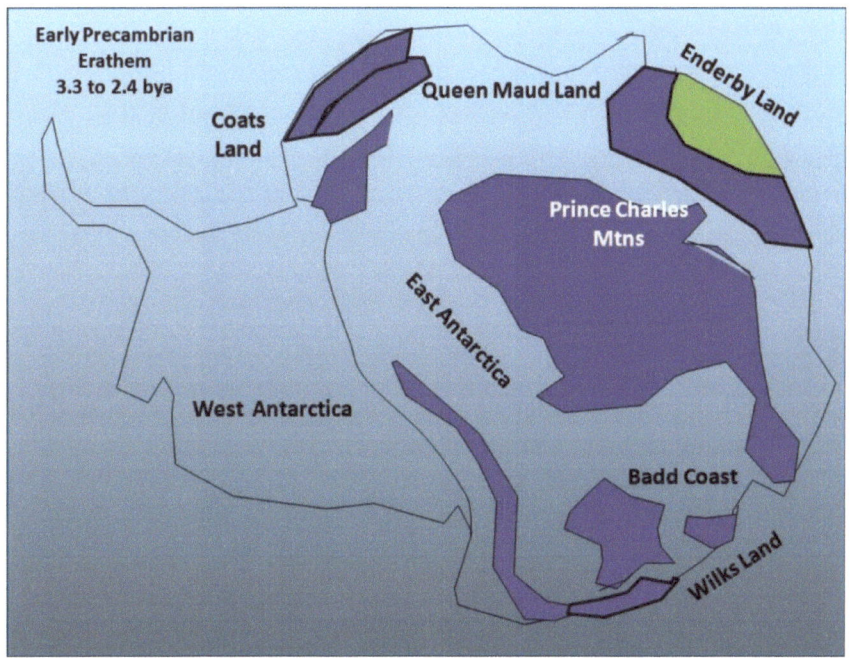

Figure 2. Intrusives and metamorphic rocks were added to East Antarctica during the Early Precambrian Era (purple).

Far western East Antarctica basement consisted of 3.0 billion year old granites beneath West Dronning Maud Land (Queen Maud Land), called the Grunehogna Craton, thought to be a fragment of the Kalahari Craton of South Africa, left attached to East Antarctica (**Figure 2**). Basement rocks were covered by shelf or platform sediments and volcanics, aged 1.1 to 1.0 billion years ago. Intrusions were injected by sills, 800 million years ago.

Enderby Land was expanded by intrusions and metamorphism, between 3.3 and 2.4 billion years ago, adding crustal mass to the Napier Complex core craton. The Mawson Block of Adelie Land is correlated to the Australian Gawler Craton. Adelie Land basement includes 3.15 to 3.05 billion year old metamorphics overlain by 2.56 to 2.45 billion year old metamorphosed sedimentary rocks.

The central part of East Antarctica was subjected to similar metamorphic and intrusive type processes, in addition to the Badd Coast-Wilkes Land cratonic fragments. The Wilkes Land fragment stretched to the northwest along the margin interior.

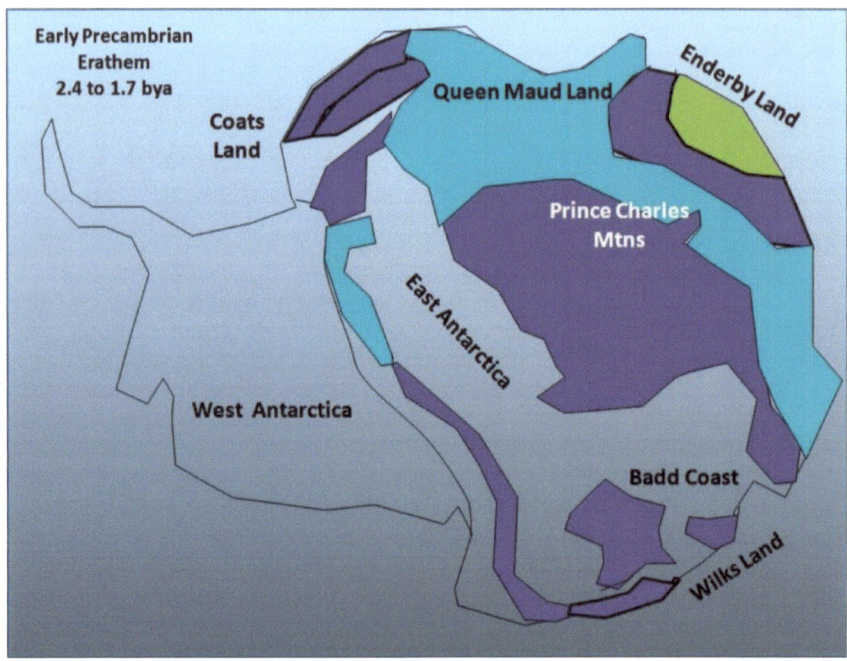

Figure 3. The Vestfold Hills (blue green) were accreted between Enderby and Queen Maud Land, filling in the East Antarctic craton.

Between 2.52 and 2.48 billion years ago, the Vestfold Hills were accreted and injected during a series of events lasting up until 1.24 billion years ago (**Figure 3**). Metamorphism occurred at depths between 16 and 24 km, rapidly brought back to within 12 km of the Earth's surface. This process suggests development probably occurred within a subduction trench which was thrusted onto the continental margin as an accretionary wedge. Dike swarms and other igneous intrusive activity followed.

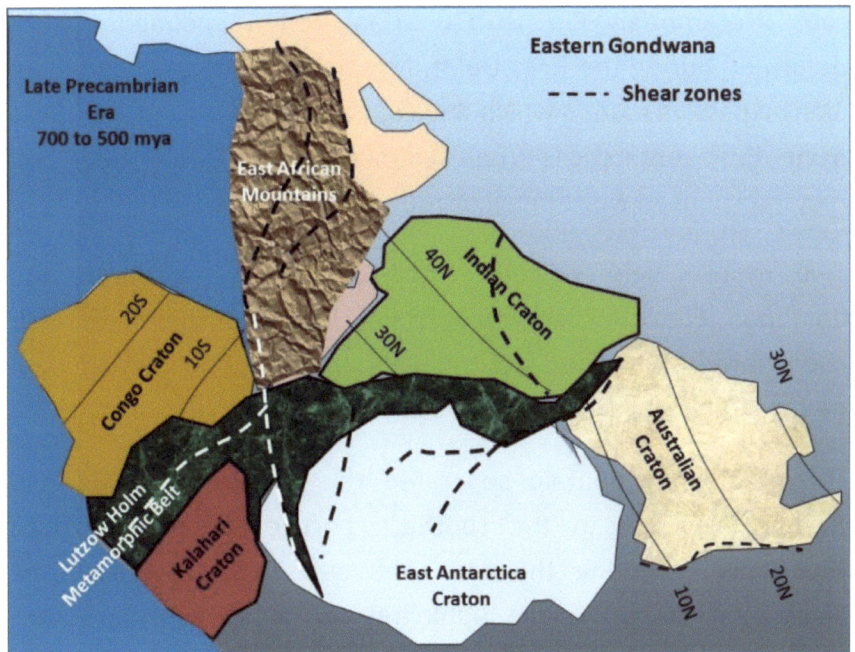

Figure 4. Closure of the Mozambique Ocean provided the mechanism for collisions between East and West Gondwana. The collision produced the East African Mountains, part of the Pan African Uplift.

When the Mozambique Ocean closed, East and West Gondwana collided and accreted together (**Figure 4**). Collisions occurred between 700 and 500 million years ago, uplifting the East African Mountains. Africa, Madagascar, Seychelles, Arabia, India, East Antarctica, South America, and Australia belonged to a single supercontinent. East Antarctica was positioned near the Equator.

East Antarctica stabilized during the Pan African Uplift. The Shackleton Mountains uplifted when South Africa collided with India along the Transantarctic Mountains. This collision was called the Ross Orogeny.

Late Precambrian Era to Cambrian Period sediments were deformed during the Ross Uplift. Marine sediments accumulated along a passive margin which developed during rifting of Laurentia from East Antarctica. The sediments were deformed and metamorphosed while being intruded by granitoids.

Two mobile belts collided with East Antarctica during the Cambrian Period. The Lutzow Holm Belt and the Prydz Belt overprinted Middle to Late Precambrian Grenville magmatic and metamorphic rocks between 550 and 515 million years ago.

Lutzow Holm Mountain separated the Grenvillian Maud and Raynor Provinces in the southern part of the East African Mountains, covering the area between East Africa and the Shackleton Range. Ocean ophiolites in the Shackleton Range support the Mozambique Ocean closure theory.

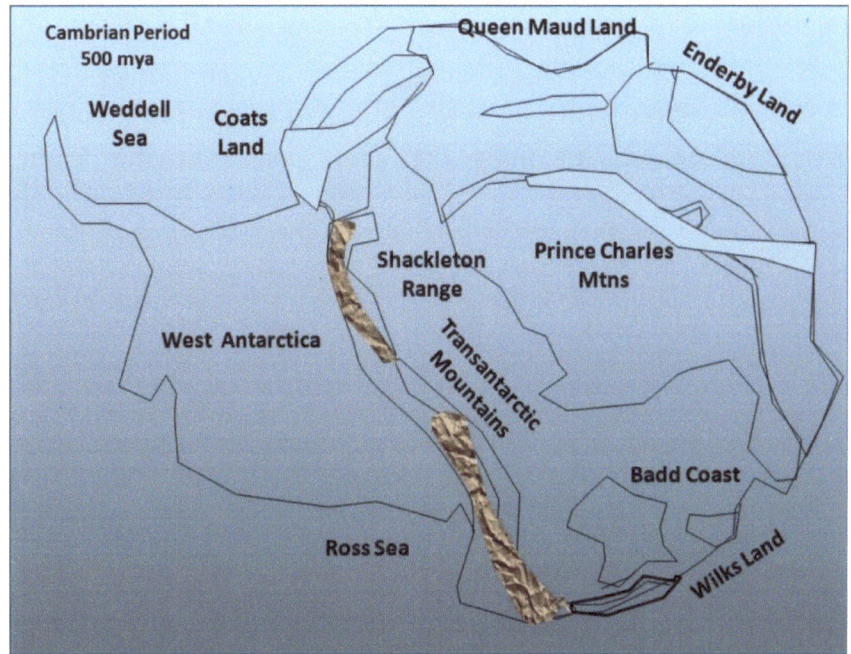

Figure 5. The Shackleton and Transantarctic Mountains uplifted and were eroded down to sea level by Ordovician time. The Ross Mountains uplifted along the same Transantarctic trend following Ordovician Period erosion.

Between the Ross and Weddell Seas, the East Antarctica margin was exposed to intermittent deformation and intrusive activity spanning from Late Precambrian into the Early Mesozoic Era (**Figure 5**).

The earliest activity began with the Beardmore Uplift which included the raising of the Transantarctic and Shackleton Ranges. The source of the uplift was an elongated sedimentary accumulation along the margin of East Antarctica. Metamorphism and igneous intrusives were covered over by Cambrian beds which covered the deformed strata.

The Beardmore Mountains were eroded over much of the Transantarctic Range. An Early Cambrian sea covered the continental shelf and interior of East Antarctica.

A carbonate platform developed, covered over by clastic and volcanics. Deformation and intrusion by granite batholiths occurred between Late Cambrian into Ordovician time, uplifting the Ross Mountains along the same trend as the former Transantarctic trend.

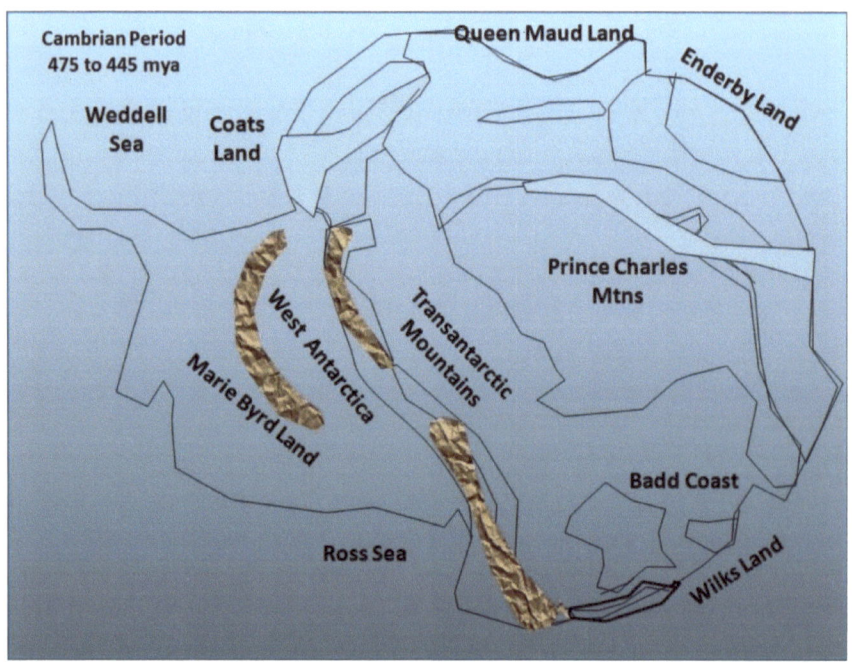

Figure 6. Mountain belts were uplifted in Marie Byrd Land, West Antarctica.

Between 475 and 445 million years ago, thick sequences of metamorphosed sediments covered Marie Byrd Land (**Figure 6**). Sediments accumulated in the same shelf-interior sea which covered East Antarctica, extending into West Antarctica. Sediments accumulated along the margins of the sea, or may have originated from the West Antarctica land mass.

Proposed theories suggest an Atlantic type margin existed during Early Paleozoic time, coupled with an island arc or continental crust collision, or associated with subduction, leaving volcanics in the Transantarctic Mountains. Later events deformed, metamorphosed, and intruded Marie Byrd Land.

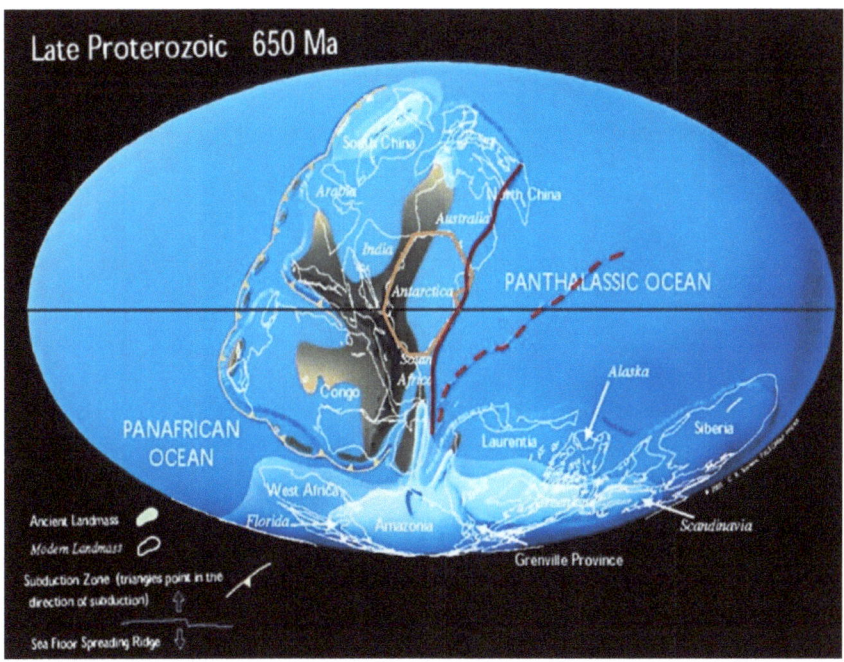

Figure 7. Antarctica was positioned at the Equator during the existence of Pannotia. A subduction trench was positioned off the eastern submerged margin, and a spreading center was located east of the trench.

Antarctica was positioned in the middle latitudes at the Equator, at the time the supercontinent called Pannotia was assembled (**Figure 7**). The western part was emergent above sea level, and the remainder was submerged beneath the western Panthalassic Ocean.

The Panthalassic subduction trench was positioned off the eastern coastline, intersecting the coastal margin in the central region. Oceanic plate subduction resulted in magmatic intrusions along the eastern margin, and within the central interior.

The oceanic lithosphere was being pushed beneath Antarctica by a spreading center positioned in the southwestern part of the ocean basin floor.

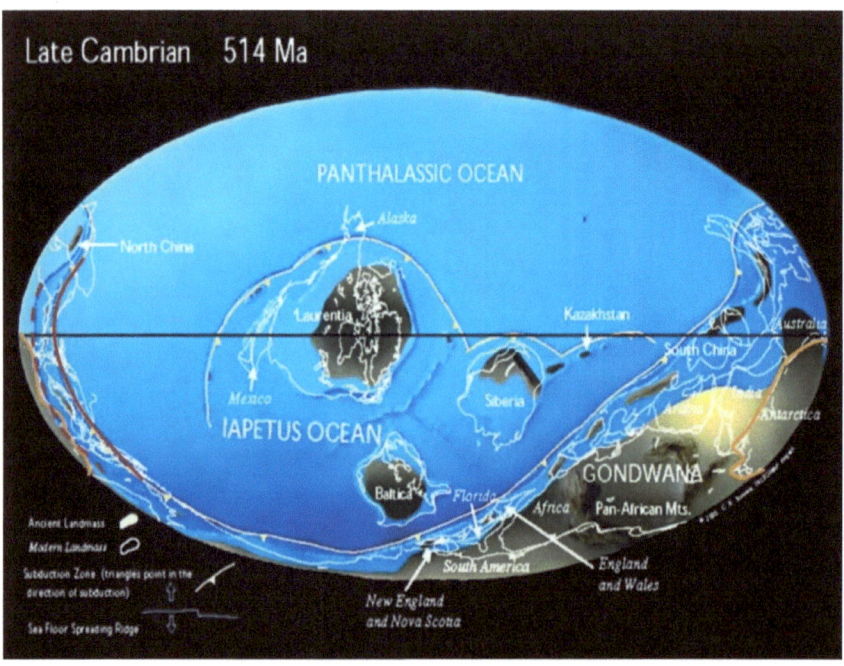

Figure 8. Gondwana was repositioned when Baltica and Siberia separated and drifted north when the Iaepetus Ocean opened. Gondwana was pushed east, and Antarctica was pushed north, trailing behind Australia.

By the Late Cambrian Period, Antarctica was positioned south of the Equator (**Figure 8**). Laurentia, Siberia, and Baltica broke away from Gondwana, drifting to the north in the Panthalassic Ocean. The Iaepetus Ocean formed between the separated smaller continents and Gondwana.

Gondwana was pushed southwards back into the lower southern latitudes. Antarctica remained positioned around 30 degrees south latitude, attached to the southern Australian margin. Antarctica was emergent above sea level.

Along the Eastern Antarctica margin, a former subduction zone was extinguished when the margin crust was pushed on top of the trench, forcing the trench to jump eastward. A new island arc system developed in the fore arc area above the subducting oceanic plate. The partial melting of the oceanic slab contributed to intrusive, deformation, and metamorphic activity in Eastern Australia.

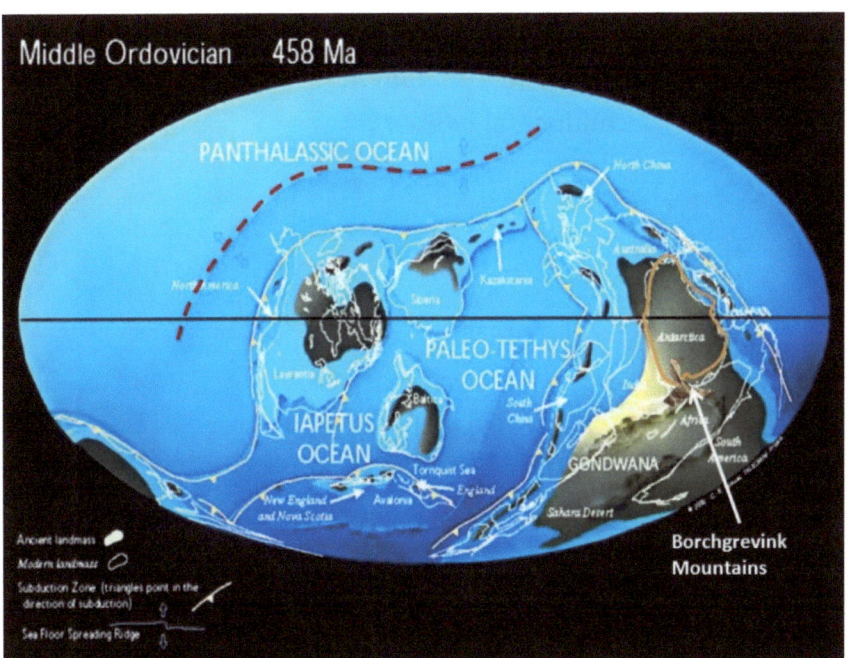

Figure 9. Antarctica was pushed north back into the Equatorial region when Gondwana continued counterclockwise rotation during the opening of the Paleo Tethys Ocean.

Antarctica was pushed back to the north by a counterclockwise rotation of Gondwana (**Figure 9**). The Paleo Tethys Ocean opened up west of Antarctica. The northern Panthalassic trench was pushed northward, and a new spreading center developed in the northern Panthalassic Ocean.

Antarctica remained emergent above sea level. The shifting Gondwana supercontinent uplifted the Pan African Mountains in Africa and South America, extending northward into southern Antarctica. The uplift developed the Borchgrevink Mountains.

Island arcs continued to erupt along the eastern fore arc basin from trench subduction of the Panthalassic Ocean plate beneath Antarctica. Subduction along the eastern trench provided the mechanism for deformation, metamorphism, and intrusive uplift of the Antarctica continental interior.

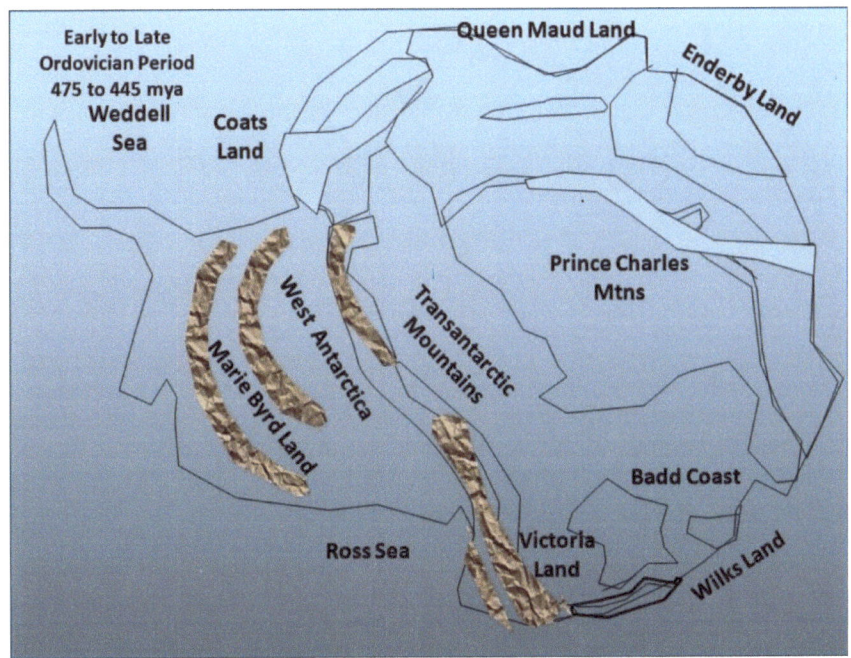

Figure 10. The Borchgrevink Mountains uplifted in Victoria Land along a northwesterly trend that ended in Marie Byrd Land.

Victoria Land was metamorphosed and intruded during the Middle Paleozoic Era (**Figure 10**). The associated uplift belonged to the Borchgrevink Mountains. Late Permian to Triassic Period sediments covered the eroded mountains.

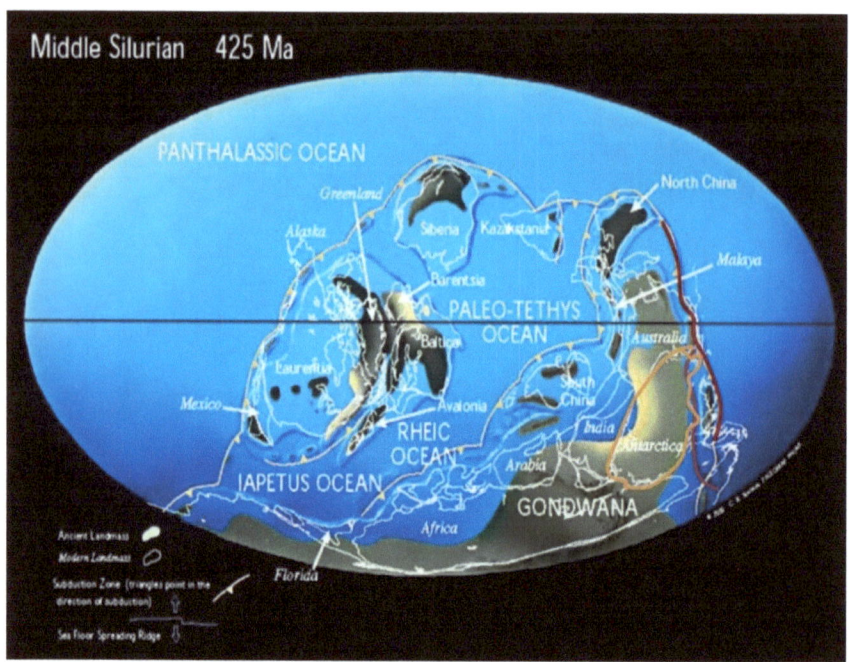

Figure 11. A subduction zone split occurred along the northeastern margin of Antarctica. The new trench subducted beneath Eastern Antarctica, injecting magmatic intrusions into continental crust.

Gondwana reversed rotation to a clockwise motion, sending Antarctica drifting south of the Equator, again (**Figure 11**). Portions of West Antarctica were invaded by a shallow sea while the majority of the continent remained above sea level.

Along the eastern margin, the rotation of Gondwana developed a new subduction trench split from the main trench system. Parts of the margin drifted above the trench. Island arcs continued erupting in the back arc basin to the new trench segment.

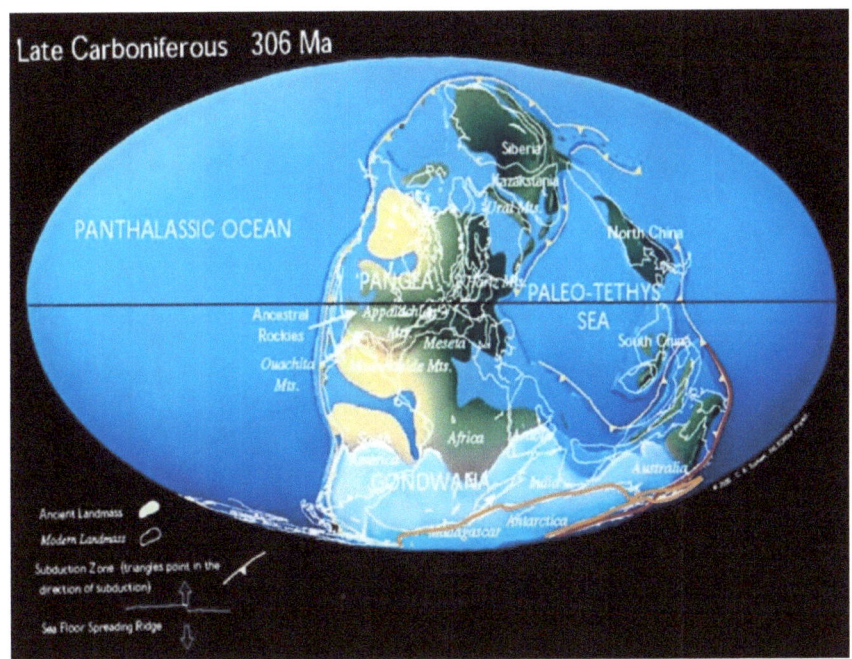

Figure 12. Gondwana rotated into the South Pole. Most of the supercontinent was covered by glaciers, including Antarctica and parts of Australia.

Continued rotation of Gondwana pushed Antarctica, India, and parts of Australia into the South Pole (**Figure 12**). Antarctica and India were entirely covered by glacial ice, including parts of Africa. Antarctica changed orientation from north-south to east and west. The eastern margin now became the southern margin.

Subduction of the Panthalassic trench continued beneath the ice sheet along the southern margin, erupting volcanics in a sub-glacial environment.

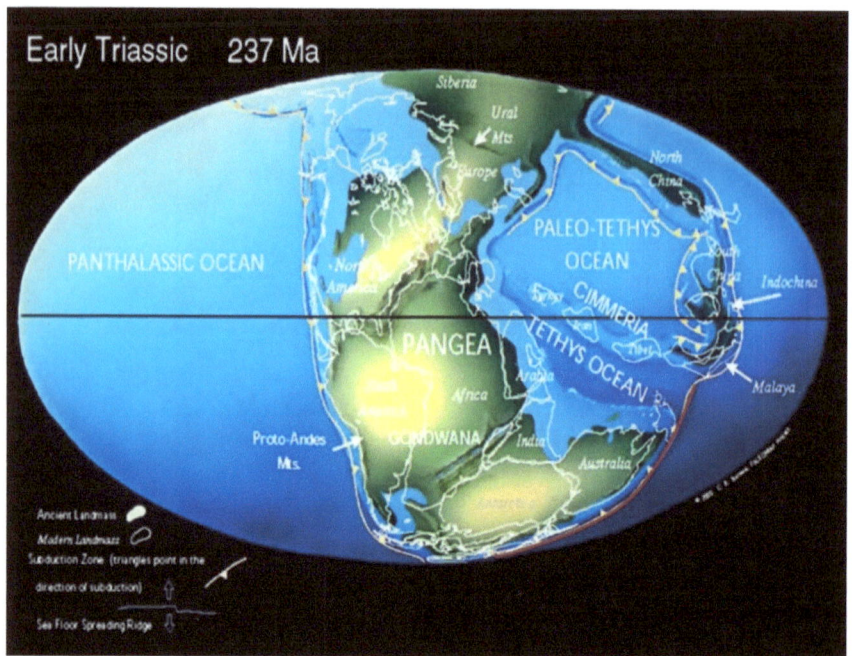

Figure 13. Antarctica drifted north away from the South Pole, melting glaciers while seas advanced across East Antarctica.

At the onset of the Mesozoic Era, the Weddell Uplift was represented by the Pensacola Mountains, developing along the East Antarctica margin (**Figure 13**). The Ellsworth Mountains uplifted within interior East Antarctica.

Antarctica moved north, out of the South Pole. Glaciers melted, leaving the continent exposed to subaerial erosion. The Panthalassic trench was positioned offshore. A fore arc basin developed between the trench and margin. A shallow sea advanced into the East Antarctica interior Ellsworth Mountain region.

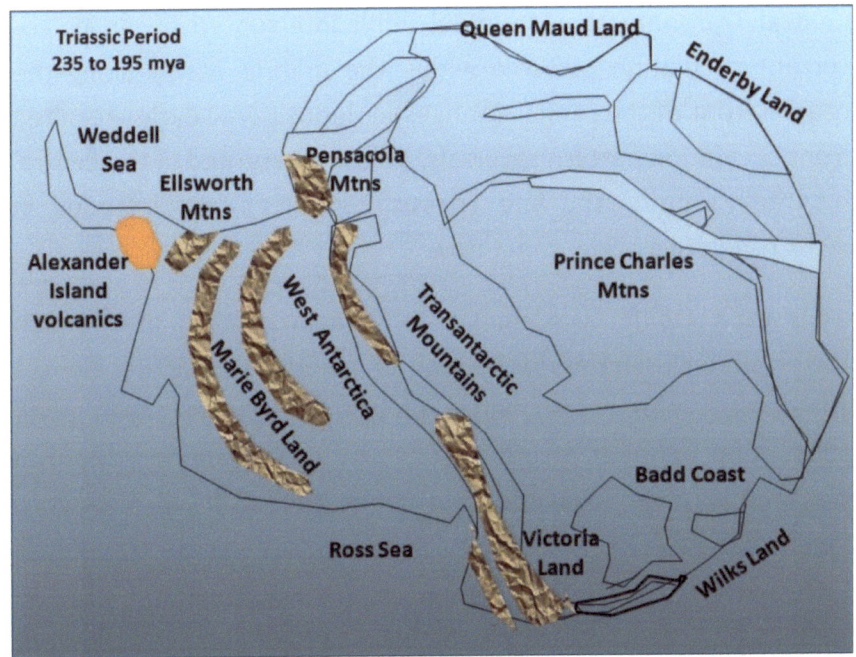

Figure 14. Gondwanan Uplift included the Pensacola Mountains, the Ellsworth Mountains, and volcanic eruptions within the Alexander Island terrane.

Depositional basins accumulated thick sedimentary deposits (**Figure 14**). Glacial sediments mixed in with marine sediments in the Ellsworth-Pensacola region when the Gondwanan Uplift began.

A regional anticline (called a geanticline) uplifted to form the Transantarctic Mountains in West Antarctica, separating marine from non-marine basin deposits. Deformation, low grade metamorphism, and intrusive activity characterized the Gondwanan Uplift. Transfer faults and grabens were common in the mountains. The mountain front was uplifted along a normal fault producing an escarpment along one half to three quarters of the entire length of the mountain front.

Central Transantarctic Mountain uplift interrupted ocean paleo-current circulation patterns within the marine basins along the coastal front of the range. Changes in depositional cycles resulted from circulation pattern reversals. Volcanics erupted in the central mountains during the late Triassic Period. Plutons intruded in West Antarctica at the same time.

Gondwana Uplift was similar to Andean type plate margin tectonism including explosive type of eruptions occurring along a continental margin magmatic arc in response to ocean plate subduction beneath a continental margin. Submarine volcanism was restricted to Alexander Island and the southern Antarctica Peninsula.

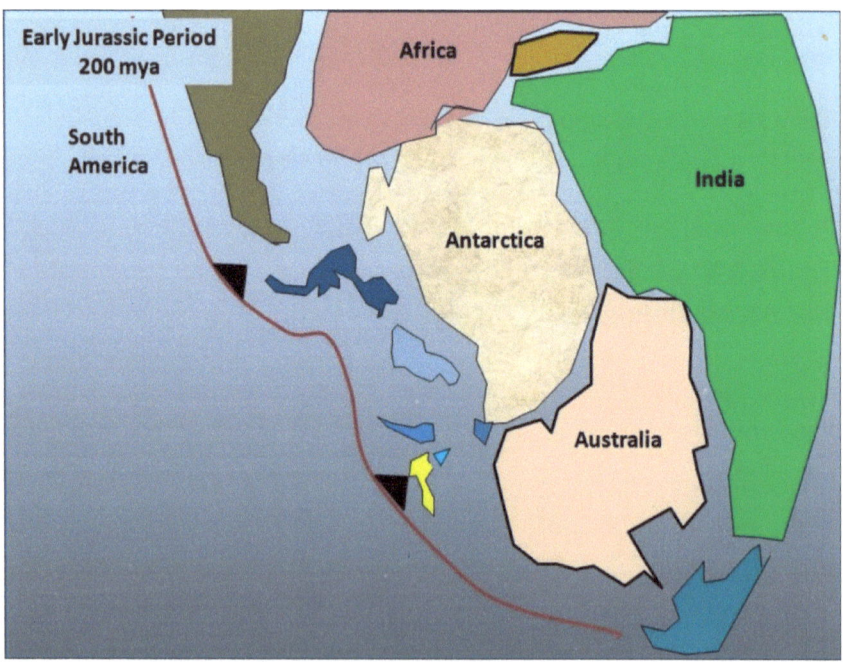

Figure 15. Antarctica was surrounded by Australia to the south, India to the east, and Africa to the north. Alexander Island (dark blue), Marie Byrd Land (lavender), New Zealand (medium blue and yellow), and Tasmania (southern tip of Antarctica) islands were present in the back arc region of the Pacific Plate trench.

The main crustal plates of West Antarctica included the Ellsworth Whitmore Mountains, Antarctic Peninsula, Thurston Island, and Marie Byrd Land (**Figure 15**). Parts of Marie Byrd Land belonged to both East and West Antarctica.

The Antarctica Andean Uplift resulted from the northward drifting of Antarctica against Africa. Depositional basins developed in South Georgia, positioned in the Scotia Sea, and in eastern Ellsworth Land during the Gondwana Uplift. The Scotia Sea occupied the northern part of the Weddell Sea basin. Metamorphism was absent, but intrusive activity was common throughout the Jurassic Period.

South Georgia was intensely deformed into overturned folds and thrust faults. Folding structures were similar to those occurring in the South American Andean Mountains. Ellsworth Land exhibited similar folding patterns.

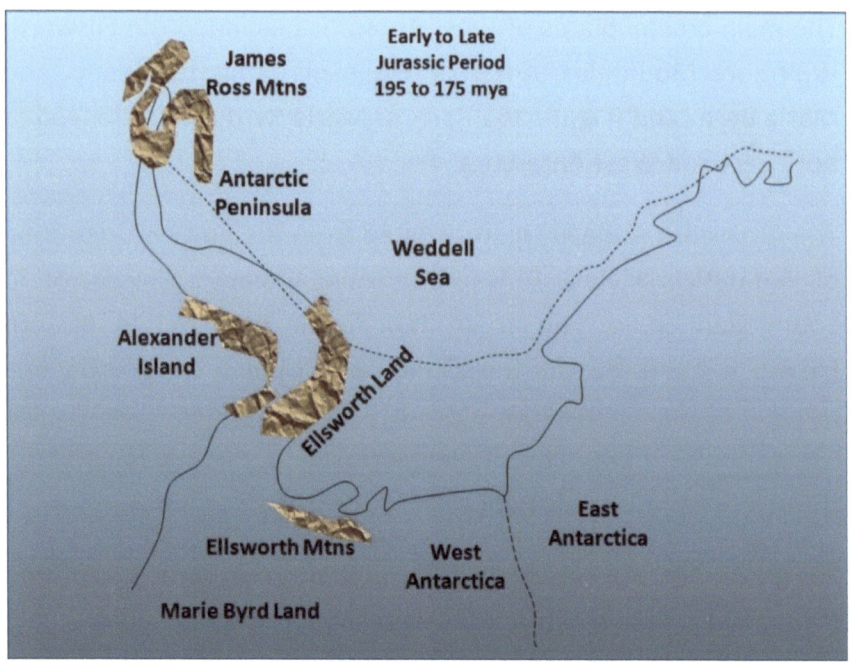

Figure 16. Gondwanan Uplifting continued in Ellsworth Land, on Alexander Island, and in the Ross Mountains at the northern tip of the Peninsula.

Back arc basins formed between an island arc and continental margin in South Georgia in response to continuing Gondwanan uplift (**Figure 16**). The Ellsworth basins developed in an intracratonic setting at a considerable distance from the continental margin.

During Late Jurassic time, volcanism erupted along an Andean type continental margin. Active subduction occurred along the West Antarctica margin resulting in plutonism occurring throughout the Cretaceous Period into Middle Cenozoic time.

Sedimentary basins developed on both sides of the rising mountain belts. Erosion of the mountains shed large volumes of continental sediments into these basins, mainly in the James Ross area and Alexander Island areas.

Southeastern Africa was likely the focus of the initial rifting where interior flood basalt eruptions began. Eruptions covered a linear belt extending from Australia (Tasman) through Antarctica (Ferrar and Queen Maud Land) into South Africa (Karoo & Ferrar) between 182 and 175 million years ago.

The basalt eruptions are thought to have occurred within the Weddell Sea basin. The basin was the location of a mega-plume which caused dome uplift and the development of a triple junction. Magma migrated along zones of crustal weakness, forming large igneous provinces. Hot plume buoyancy acting on a down going subduction slab resulted in slab flattening beneath the continental crust, the source of fissure type eruptions. The Gondwana Fold Belt deformed during uplift above the plume, beginning the rifting process.

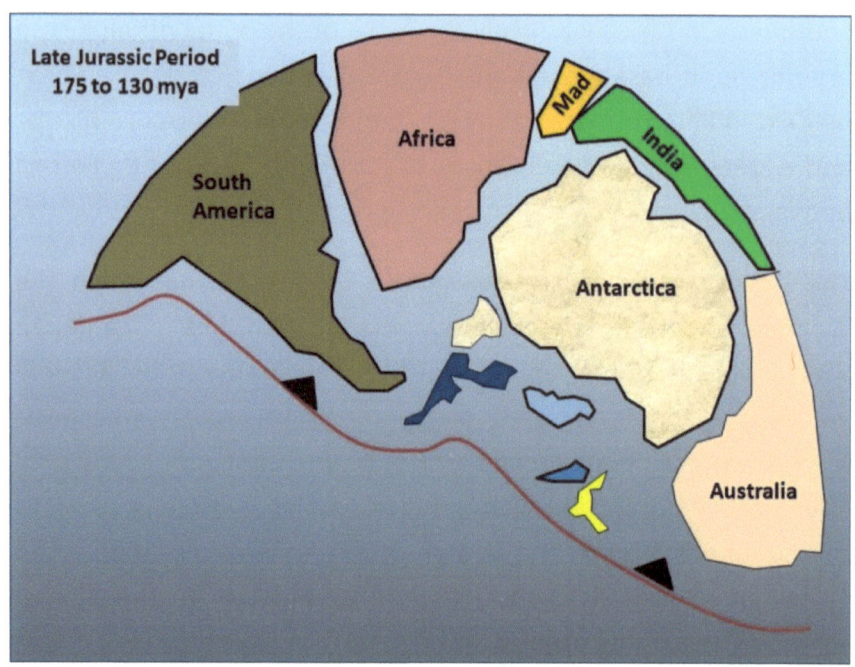

Figure 17. Continued subduction started rearranging the Peninsula terranes. New Zealand started to collide in the south. Marie Byrd Land drifted closer to Antarctica, and Alexander Island was about to collide against the Ellsworth Whitmore terrane fragment which separated from the mainland.

Sill intrusions into the Transantarctic Mountains occurred about 177 million years ago into the basement and sedimentary cover rocks at the same time extrusive volcanics erupted at the surface (**Figure 17**). The mountain range consisted of a simple tilt block structure exposing Kirkpatrick Basalts along the inland flank and deeper basement exposures along the coastal flank.

Rotation of the Weddell Sea basin carried the Ellsworth Whitmore Mountains from southeastern Africa to its present day position. Completion of West Antarctica rotation ended 165 million years ago. Rotation was accomplished by fault displacement occurring beneath basin floor sediments, as opposed to oceanic crust formation.

A back arc basin developed along the Antarctic margin from clockwise rotation of the Weddell Sea during Late Jurassic time.

Gondwana began to rift apart during the Late Jurassic Period, 160 million years ago along a right lateral translational extension zone when East Antarctica (Antarctica, Australia, India, and New Zealand) separated from West Antarctica (South America and Africa) starting in the north, separating to the south. Magmatic intrusion, rotation, and displacement of micro-plates accompanied the initial breakup.

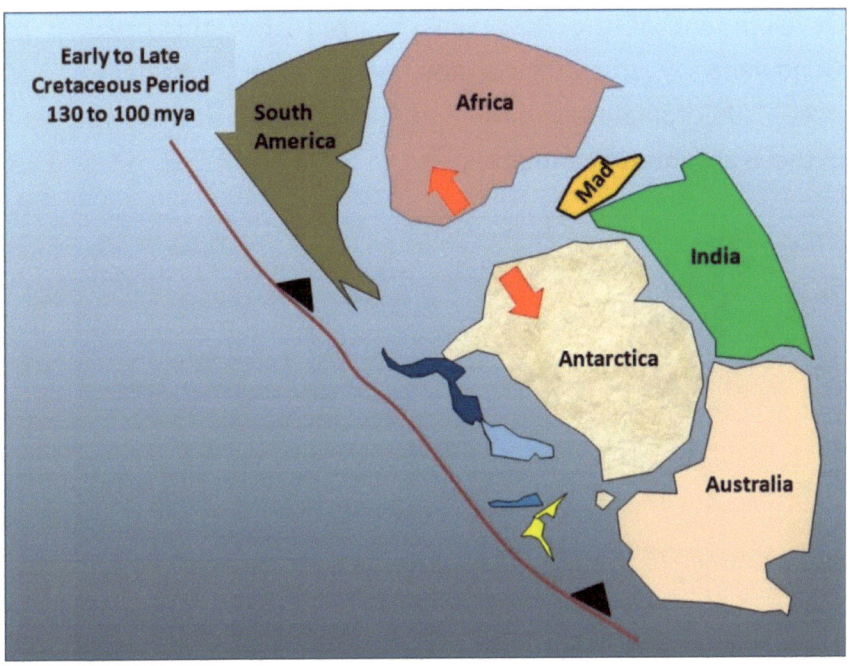

Figure 18. Development of the West Antarctica Rift System began when Gondwana broke apart during separation of Antarctica from Africa. India and Australia remained attached to Antarctica. Assembly of the Peninsula region began. Ellsworth Whitmore reattached to East Antarctica.

Early Cretaceous breakup of Gondwana continued (**Figure 18**). Stress directions changed from north to south to an east to west direction. The two plate system of East and West Antarctica was replaced by a multi-plate system.

The Transantarctic Mountains were buried and exhumed in three stages. The first two stages occurred in the Early and Late Cretaceous Period between 125 and 110 million years ago when the Admiralty Mountains uplifted during crustal stretching between Antarctica and Australia.

Development of the Ross Sea occurred after the breakup of Gondwana by crustal extension. Magmatism, subsidence, and possibly lateral faulting in the Ross Sea occurred at the same time in the Transantarctic Mountains.

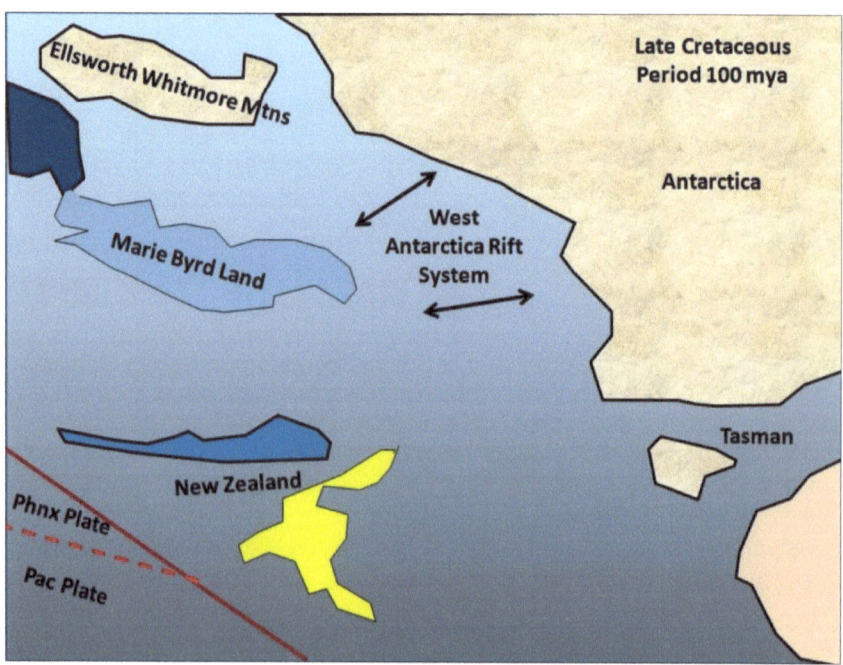

Figure 19. The West Antarctica terranes continued drifting away from East Antarctica when the West Antarctica Rift System continued opening. Ellsworth Whitmore was separated from the continent again.

Marie Byrd Land began as a subduction zone complex during most of the Paleozoic and Mesozoic Era (**Figure 19**). During the Late Cretaceous Period, subduction processes transformed into extension processes in New Zealand, Marie Byrd Land, and within the Ross Embayment. Sea floor spreading separated the Campbell Plateau from Marie Byrd Land, 84 million years ago.

A failed rift developed when initial seafloor spreading between Antarctica and Australia began 95 million years ago, forming the Rennick Graben. The Tasman terrane separated from East Antarctica, and the Phoenix Plate separated from the Pacific Plate along a new spreading ridge center which developed west of the New Zealand islands.

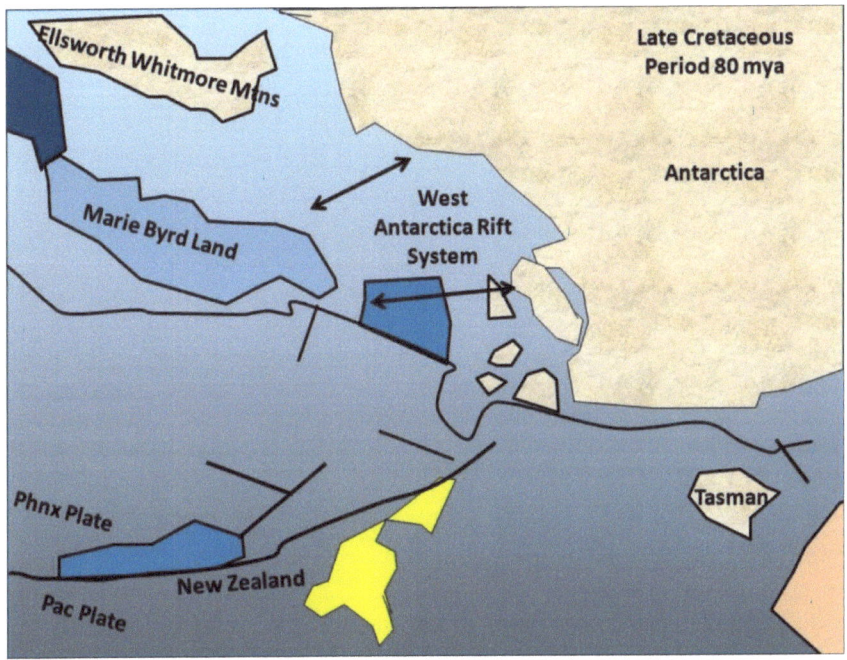

Figure 19. Marie Byrd Land continued to drift away from East Antarctica. Magmatic intrusion occurred in response to West Antarctica rifting. The Phoenix Plate separated from the Pacific Plate, slicing through North (yellow) and South (medium blue) New Zealand.

Late Cretaceous rifting between East and West Antarctica occurred during subduction of the Pacific-Phoenix spreading center ridge along the Pacific margin outboard of New Zealand, extending into Marie Byrd Land (**Figure 19**).

Extension occurred within New Zealand, coupled with a change from subduction to rift magmatism in Marie Byrd Land, around 100 million years ago. Divergent motion between the Pacific and Antarctic plates transferred New Zealand from the Antarctic plate to the Pacific plate when a new spreading ridge separated the two plates from each other.

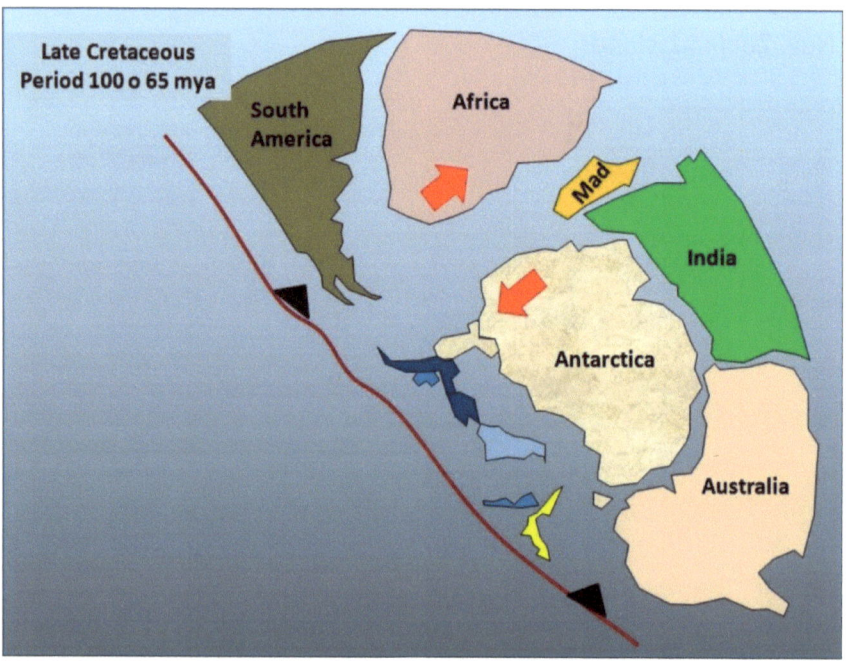

Figure 20. Rifting converted into shearing between Africa and Antarctica plates. Peninsula Antarctica began assembling when Ellsworth Whitmore collided with East Antarctica, and the Alexander Island terrane collided against Ellsworth Whitmore.

Around 110 million years ago, West Antarctica rotated close to its present day position and configuration with East Antarctica (**Figure 20**).

India and Antarctica began to rift apart. Extension between Australia and Antarctica began. Sea floor spreading started around 95 million years ago.

Between 105 and 85 million years ago, extension accompanied crustal thinning in the West Antarctica Rift System opening between East and West Antarctica. Victoria Land began to separate after the Middle Oligocene Epoch. Detachment faults appeared to be the mechanism which caused extension to occur between East and West Antarctica, around 103 million years ago.

Detachment faults develop when blocks of rocks slide along gravity faults to a new position where older rocks occur on top of younger rocks. Detachment faults may be misinterpreted as low angle thrust faults.

By Late Cretaceous time, Antarctica rotated into the South Polar Region. The breakup of Gondwana was completed 84 million years ago.

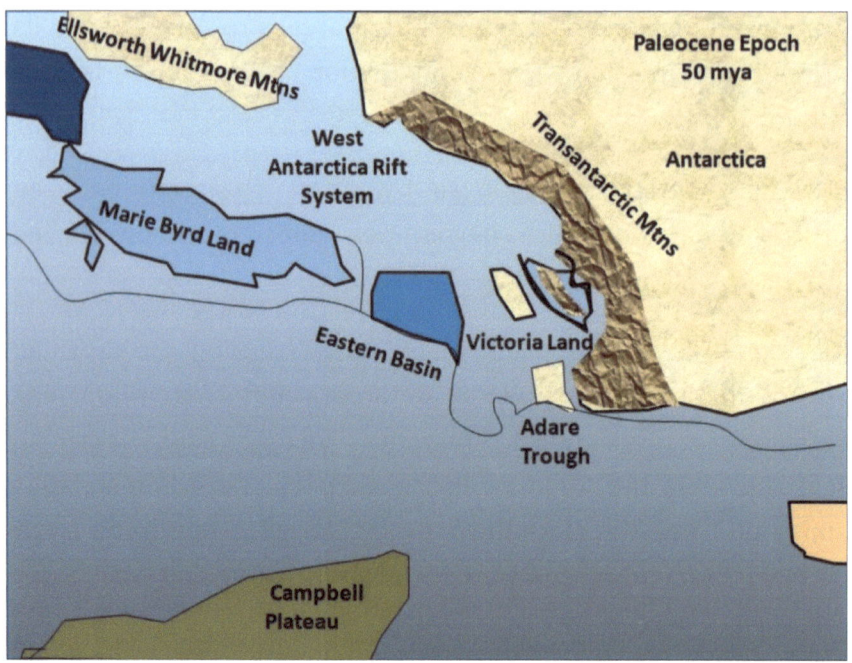

Figure 21. Sea floor spreading west of Victoria Land developed the Adare Trough, initiating rifting between East and West Antarctica. The Transantarctic Mountains began to uplift.

At the beginning of the Paleocene Epoch, sea floor spreading in the Adare Trough extended southward into the Victoria Land Basin, along the Transantarctic Mountain front (**Figure 21**). The Adare spreading center initiated rifting between East and West Antarctica, along the West Antarctica Rift System, triggering Transantarctic Mountain uplift.

Perpendicular faulting to the general mountain range trend displaced and dissected the range along normal faults, vertically. Shear faults converted into left lateral faults around 30 million years ago, ending Adare seafloor spreading.

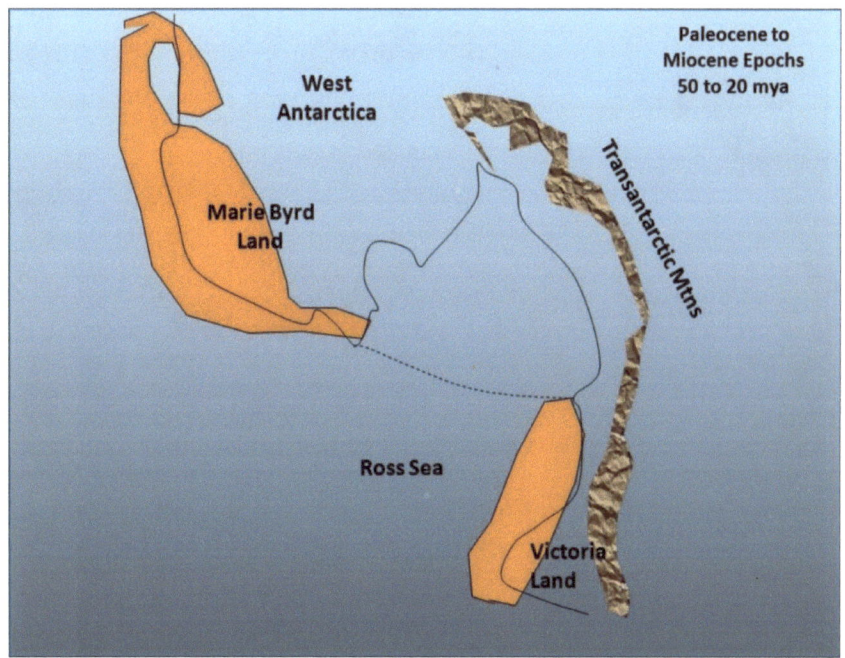

Figure 22. Basalt eruptions in Marie Byrd Land and in Victoria Land occurred along fissures connecting the upper mantle to the surface crust. Transantarctic Mountain uplift continues at present.

The South Sandwich Islands represent the early stages of island arc development, resulting in the opening of the Scotia Sea basin (**Figure 22**). The Drake Passage was the location of an active spreading center, 20 million years ago, presently inactive. A southeasterly dipping subduction zone developed the South Shetland Islands.

Volcanism in the Deception Islands originated from the upper mantle, along with West Antarctica basalt volcanics. Antarctica appears to be undergoing a period of rifting at the present time. Uplift along the Transantarctic Mountains continues.

Rapid spreading continued between Antarctica and Australia, 45 to 28 million years ago. Rifting was limited to the Terror Rift within Victoria Land, western side of the Ross Embayment.

Figure 23. The Antarctic continent was covered in ice again, 15 million years ago. Ice sheets continue to cover the continent. Climate changes occurred around the world since glaciers buried Antarctica.

Erosion of the Transantarctic Mountains resulted in scarp retreat, formation of planation surfaces, and down cutting by river systems up until 15 million years ago (**Figure 23**). Present day landscapes exhibit river erosion, transport, and deposition modified by glacial advance and retreat cycles. Tectonic activity continues within the Transantarctic Mountains. Volcanism continues in the McMurdo region.

References

Chapter I: Google Earth Nations Online Project. 2014. Australian satellite image was extracted from the World Satellite Map posted on the internet by Google Earth.

Huston, D.L., Blewett, R.S., and Champion, D.C. 2012. Australia through time: a summary of its tectonic and metallogenic evolution. Geoscience Australia. Australian crustal elements and basin map series redrawn to depict historical progress in the evolution of Australian cratonic elements; tectonic maps models redrawn for texture and color enhancements; text extracted and edited for introductory presentation.

Meert, J.G., and others, 2010. Precambrian crustal evolution of Peninsular India: a 3.0 billion year odessy. Journal of Asian Earth Sciences. Rodinian and Columbian supercontinent maps were redrawn from figures presented in the publication.

Scotese, C.R. 2013. Copyright. PALEOMAP Project. Paleogeographic globe maps used with permission.

Chapter II: Elliot, D. H. 1973. Tectonics of Antarctica: A review. American Journal of Science, Vol. 275-A, p. 45-106. Line maps showing mountain belts redrawn from the set of maps contained within the publication.

Fitzgerald, P. 2002. Tectonics and landscape evolution of the Antarctica plate since the breakup of Gondwana, with an emphasis on the West Antarctica Rift System and the Transantarctic Mountains. Royal Society of New Zealand Bulletin 35; 2002; 453-469; 1-877264-06-7. Mesozoic Era paleogeographic maps redrawn to add in color and texture. Text extracted for Mesozoic Era geologic history. West Antarctica Rift Zone paleo-reconstruction maps redrawn to add in color and texture.

Grikurov, G.E., Mikhalskii, E.V. 2002. Tectonic structure and evolution of East Antarctica in the light of knowledge about supercontinents. Russian Journal of Earth Sciences, Vol. 4, No. 4, pp. 247 to 257. Portions of Precambrian Era geologic history text extracted and edited.

Harley, S.L. ___. The Geology of Antarctica. Geology, Vol. IV. Encyclopedia of Life Support Systems. Text and maps extracted to present Early Precambrian Era tectonic history.

Hambrey, M. ___. Geology of Antarctica. Aberystwyth University, Wales, UK. Text extracted and revised to provide an introductory summary of Antarctica geologic history.

Meert, J. G., and others. 2010. Precambrian crustal evolution of India: a 3 billion year odessy. Journal of Asian Earth Sciences. Columbia supercontinent map redrawn to add color and textures.

NASA, ___. Satellite image of Antarctica digitally modified to remove present day ice cap.

Scotese, C.R. 2013. Copyright, PALEOMAP Project. Paleogeographic globe maps used with permission.

Wikipedia, Online Encyclopedia, ___. East Antarctica Shield. Published on the internet. Precambrian Era text extracted from numerous sources referenced online.

Other Titles Available in this book series

Introduction to Global Plate Tectonics I: Plate Tectonics Theory, Paleogeography, and the Ocean Basins.

Book I presents the basic theory for understanding the driving forces behind continental drift, and the opening and closing of the ocean basins. This is the first of a series of ebooks which offers interested individuals and students a non-technical approach towards introducing *plate tectonics theory*, and the mechanisms for understanding the distribution of continents around the globe. *Paleogeography* is the study of reconstructing continental positions throughout geologic time, up to the present positions. Two maps are offered for predicting future continental positions as they may appear tens of millions years from now. The *Ocean Basins* presents the geologic history of the major oceans and how they formed as continents drifted apart, collided, and reassembled throughout geologic time. Video slide shows are available for viewing on Slideshare.com (free of charge). Enter search word Category Education, keyword plate tectonics. The slide show offers basic animation, audio text narrative, and music soundtrack. List Price: $24.18. Available from Amazon.com.

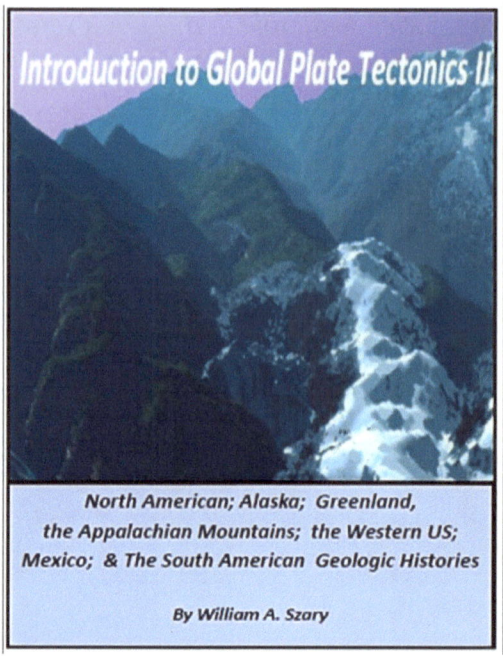

North American; Alaska; Greenland,
the Appalachian Mountains; the Western US;
Mexico; & The South American Geologic Histories

By William A. Szary

Introduction To Global Plate Tectonics II: North American Tectonic History

Books II presents the Precambrian Craton & Phanerozic Geologic History of the North American craton from Early Precambrian time, greater than 2 billion years ago, to the Early Paleozoic Era Cambrian Period, 530 million years ago. The tectonic assembly of Alaska, the Appalachian Mountains, the Western US, Greenland, Mexico, and South America are discussed. The book utilizes many picture maps, accompanied by text discussion, to present the subject matter in an understandable presentation. Video slide shows are available for viewing on Slideshare.com (free of charge), Category Education, keyword plate tectonics. The slide show offers basic animation, audio text narrative, and music soundtrack for the North American Precambrian and Phanerozoic Eras. South American, Alaska, Mexico, and Greenland are not included with the presentation. Book List Price: $23.13. Available from Amazon.com.

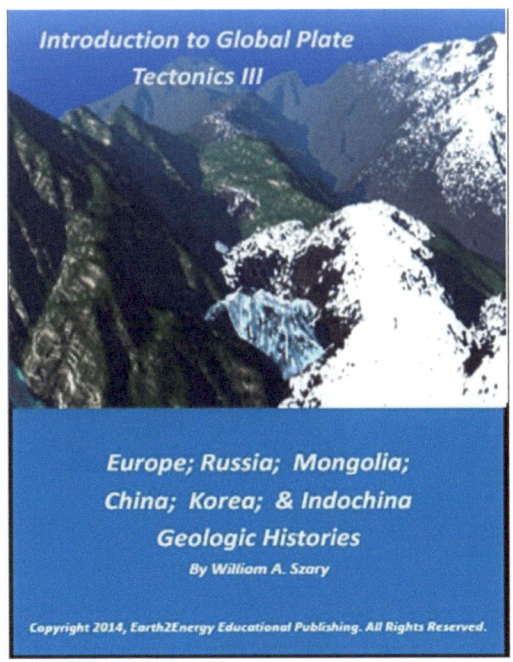

Introduction To Global Plate Tectonics III: Europe, Russia, Mongolia, China, Korea, and Indochina is the third part of a five part series covering the subject of plate tectonics, paleogeography, and the drifting and build out of continents. Each book was designed as a picture guide containing many images extracted from scientific journal articles written by research professors on the subject matter. The text content has been rewritten to help explain technical terms, converting the terms into a more understandable language for those interested in learning about geological sciences, but have not yet mastered the terminology. Images were redrawn by adding color and texture to highlight key events explained by the text.

Part III covers the development of the European basement continuing through to the present day. Chapters on Russia, Mongolia, China, Korea, and Indochina geologic histories are included Book III. This book focuses on the more technical aspects of continental development. Book List Price: $34.33.

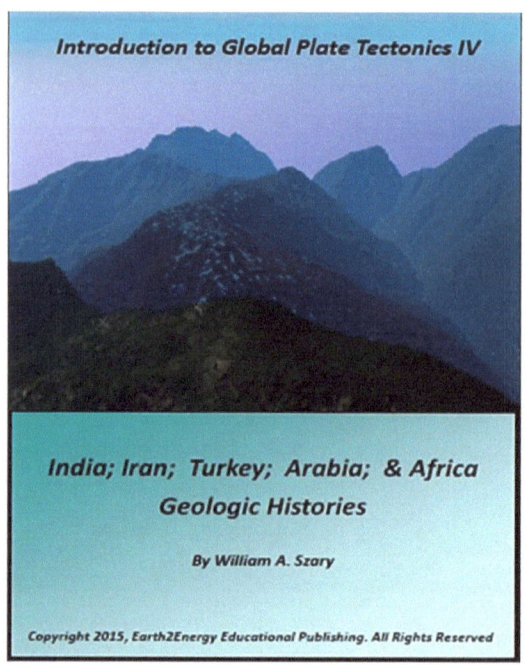

Introduction to Global Plate Tectonics IV
India; Iran; Turkey; Arabia; & Africa
Geologic Histories
By William A. Szary

Introduction to Global Plate Tectonics IV: Tectonic Histories of India, Iran, Arabia & Africa. Introduction To Global Plate Tectonics IV is the fourth part of a five part series covering the subject of plate tectonics, paleogeography, and the drifting and build out of continents. Each book was designed as a picture guide containing many images extracted from scientific journal articles written by research professors on the subject matter. The text content has been rewritten to help explain technical terms, converting the terms into a more understandable language for those interested in learning about geological sciences, but have not yet mastered the terminology. Images were redrawn by adding color and texture to highlight key events explained by the text.

Part IV covers the development of the Indian, Iranian, Turkish, Arabian, and African basement continuing through to the present day. This book focuses on the more technical aspects of continental development. Book List Price: $40.63.

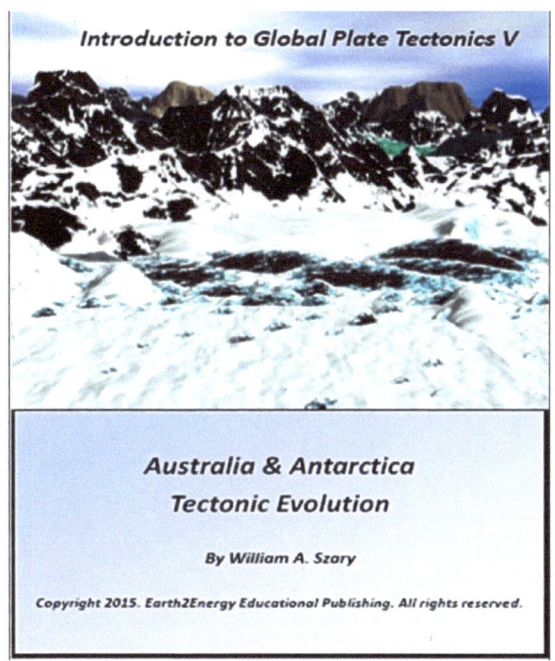

Introduction to Global Plate Tectonics V: Australia and Antarctica. Introduction To Global Plate Tectonics V is the fifth part of a five part series covering the subject of plate tectonics, paleogeography, and the drifting and build out of continents. Each book was designed as a picture guide containing many images extracted from scientific journal articles written by research professors on the subject matter. The text content has been rewritten to help explain technical terms, converting the terms into a more understandable language for those interested in learning about geological sciences, but have not yet mastered the terminology. Images were redrawn by adding color and texture to highlight key events explained by the text.

Part V covers the development of the Australia and Antarctica Precambrian Era basement continuing through to the present day. This book focuses on the more technical aspects of continental development. Book List Price: $16.48. Available from Amazon.com.

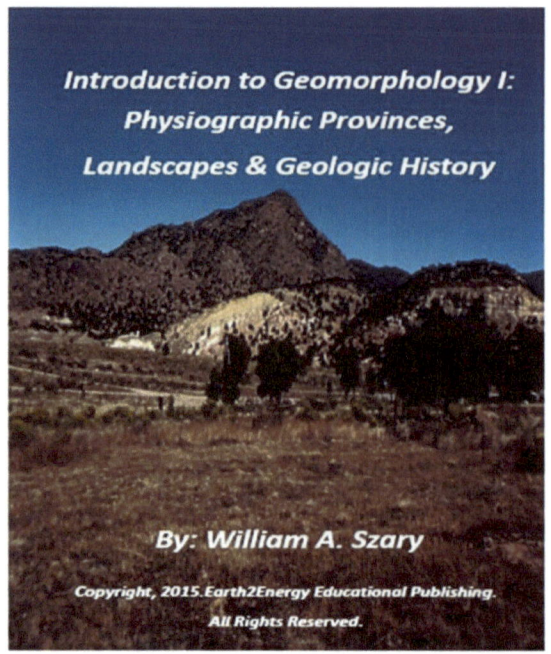

Landscapes and landforms are expressions of geologic structure, processes, and time. Each landscape expresses certain physical similarities grouped into provinces based on geologic history, geologic structure, rock formations, and the processes which weather and erode landscapes based on climatic conditions. All factors play a role in shaping and sculpting landscapes and landforms.

Introduction to Geomorphology I presents a review of the geologic history behind each physiographic province, providing typical and atypical photographic representations for each province recognized in the continental United States (125 photographs).

Introduction to Geomorphology II presents structures, landforms, and geologic processes presenting constructional (tectonic and volcanic landforms), destruction (weathering processes), erosional processes (mass wasting, hill slope evolution), fluvial (river), and coastal processes using many photographic examples for each subject presented. Book List Price: $34.68.

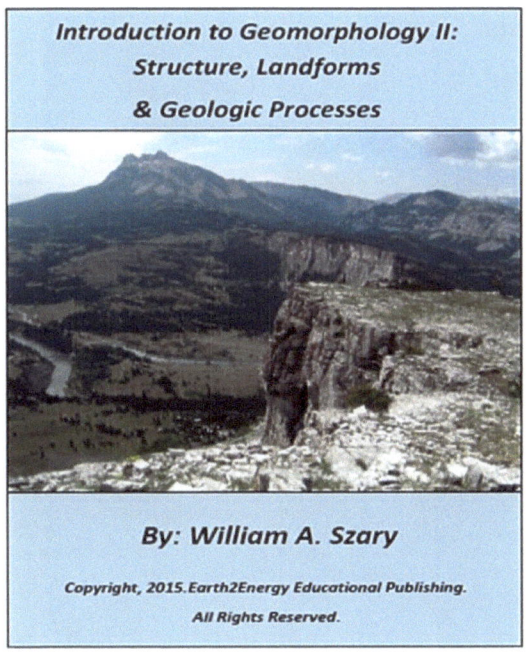

Introduction to Geomorphology II: Structure, Landforms & Geologic Processes

By: William A. Szary

Book II of the geomorphology series continues with expanding the description of geomorphic provinces describing landscapes in the context of geologic structure, landforms, and the basic principles of processes which shape landforms and landscapes. Book I addressed the geomorphic provinces in terms of the geomorphic province, landscapes and geologic history behind each province.

Many photographs are presented in this book covering constructive, destructive, mass wasting, fluvial (river) processes, and coastal processes. Photographs that were obtained from the public domain sources are cited for the source and author. Where no citations are provided, photographs were taken by the author.

Both books in this series are intended for those interested in earth sciences at the secondary school, community college, and first year undergraduate level of study. Book List Price: not yet determined.

For additional information contact William Szary at wszary@netzero.net.

Earth2Energy Educational Publishing
Tampa FL 33618
Earth2Energy is a registered trademark.

www.ingramcontent.com/pod-product-compliance
Lightning Source LLC
Chambersburg PA
CBHW040809200526
45159CB00022B/117